DR. PIERRE F. WALTER

THE SECRET SCIENCE

The Huna Knowledge of the Cosmic Energy Field

"Articles Series"

©2011 Pierre F. Walter. All rights reserved.

Published by Sirius-C Media Galaxy LLC

http://sirius-c-publishing.com

http://siriuscmedia.com

http://ipublica.com

ISBN 978-1-468137-25-5

Contact Information Dr. Pierre F. Walter

publisher@sirius-c-publishing.com

About Dr. Pierre F. Walter

http://drpfw.info

Quotation Suggestion

Pierre F. Walter, *The Secret Science: The Huna Knowledge about the Cosmic Energy Field*, Sirius-C Media Galaxy LLC, 2011

About the Author

Pierre F. Walter is an author, international lawyer, researcher, corporate trainer, and lecturer. After finalizing studies in German Law, International Law and *European integration* with diplomas obtained in 1981 through 1983, he graduated in December 1987 at the Law Faculty of the University of Geneva as *Docteur en Droit* in international law.

The doctorate was funded by scholarships from the *Swiss Institute of Comparative Law*, Lausanne, and from the *University of Geneva*, as well as a Fulbright Travel Grant for an assistantship with Professor Louis B. Sohn at *UGA Law School Department of International Law*, Athens, Georgia, USA, in 1985. Pierre F. Walter also served as a research assistant to *Freshfields, Bruckhaus, Deringer*, Cologne, Germany in 1983 and to *Lalive Lawyers*, Geneva, in 1987.

Pierre F. Walter writes and lectures in English, German and French languages; he has written *more than ten thousand pages* embracing all literary genres, including *novels, short stories, film scripts, essays, selfhelp books, monographs* and extended *book reviews*. Also a pianist and composer, he has realized 40 CDs with *jazz, newage* and *relaxation music*.

Pierre F. Walter's professional publications span the domains *International Law, Criminal Law, Holistic Science, Psychology, Education, Shamanism, Ecology, Spirituality, Quantum Physics, Systems Theory, Natural Healing, Peace Research, Personal Growth, Selfhelp* and *Consciousness Research*. 110 Book Reviews, thirty-eight audio books and more than hundred video lectures were realized in the years 2005-2010. Besides, Pierre F. Walter is author and editor of *Great Minds Series*, which features scientists, artists and authors of genius from Leonardo to Fritjof Capra.

Pierre F. Walter publishes via his Delaware firm *Sirius-C Media Galaxy LLC* and the imprints IPUBLICA and Sirius-C Media (SCM).

For the Kahunas

CONTENTS

INTRODUCTION 8
Why Studying Huna?

The Heretic Knowledge Tradition 21
Rediscovering the Real Science

- Carl-Gustav Jung (1875-1961)
- Paracelsus (1493-1541)
- Swedenborg (1688-1772)
- Mesmer (1734-1815)
- Reichenbach (1788-1869)
- Reich (1897-1957)
- Lakhovsky (1869-1942)
- Burr (1889-1973)

The Secret Science 46
The Huna Knowledge of the Cosmic Energy Field

GLOSSARY 61
Contextual Glossary

- Complexity   61
- Consciousness   62

Bibliography

Direct Perception   64

E-Force   65

Emonics   65

Definition

Emonics Vocabulary

E and E-Force

Emonic and Demonic

Emonic Charge

Emonic Awareness

Emonic Current or Emonic Flow

Emonic Integrity

Emonic Sanity

Emonic Sanity in Relationships With Children

Emonic Setup

Emonic Vibration

Emotional Flow   70

Quantum Physics   71

Definition

The Uncertainty Principle

Nonlocality

The Uncertainty Principle

What the Bleep Do We Know?!

Sadism	80
Shamanism	82
Definition	
Entheogens	

BIBLIOGRAPHY 86
General Bibliography

FROM THE SAME AUTHOR 193
A Bibliography

INTRODUCTION

Why Studying Huna?

In other productions I have elucidated in some detail the perennial research on the *cosmic life energy* as it is to be found both in the Eastern and Western science traditions.

As the purpose of each production is unique, I cannot avoid repeating myself when presenting the information, and the research, to their different audiences. And the audiences are different here, to be sure.

Emotional Flow, the first audio book where I present my research on the Oriental view of the universe as an energy-driven and *flowing* cosmos, addresses a lay public and a general science audience.[1] This production was largely inspired by my research on *Feng Shui*, the age-old Chinese science of *energy-based vibrational correspondences* in nature and human habitat.[2] To my amazement, I found as a consequence of this

---

[1] See Pierre F. Walter, *Emotional Flow*, Audio Book (2010).

[2] See Pierre F. Walter, *Basics of Feng Shui*, Scholarly Article (2011).

research that Wilhelm Reich's discovery of the *orgone* was something not as queer and outlandish as his contemporaries, and even many of our present-day Cartesian thinkers judged it to be, but an event that had to happen as a matter of *systemic logic*.[3] Hence, with great enthusiasm and an infinite admiration for Reich's courage and his stout scientific honesty I produced an audio book on Reich, entitled *Reich's Greatest Discoveries (2010)*, as well as my more specialized audio book on Reich's bioenergy treatment approach, *Orgonomy and Schizophrenia (2010)*.

This research eventually culminated in *Emonics (2010)*, a quite unusual audio book in which I present a complete and practically applicable vocabulary and terminology that could serve as a first step in the creation of a *holistic treatment approach* for sexual paraphilias. This study, which is perhaps my greatest scientific achievement so far is admittedly also perhaps my hardest-to-read production because it's really heterogeneous, an amalgam of different sources of input- This audio book may also be accessible, from its very purpose, to a rather limited audience of bioenergy researchers and forensic experts, as well as systemically trained psychiatrists. I cannot really imagine *Emonics* to becoming once a choice for the lay reader.

And this is exactly the reason for the present scholarly article to be born. I have recently modified the purpose and potential audience of my ever first web site, *Autohuna* dot

---

3 On the matter of systems theory, the science of living systems, see, for example, Ludwig von Bertalanffy, *General Systems Theory (1976)* and Fritjof Capra, *The Web of Life (1997)* and *The Hidden Connections (2002)*.

com, created back in 1996, which initially was quite a clutter box of essays on new spirituality.[4] After a while, I began to reflect if I was really following my initial intention for that site. Let us have a look at its name, for it is quite an extraordinary matter. *Autohuna* is a term not born, and not rooted, in Western civilization but an expression created by the Tsalagi (Cherokee) nation of North-American natives. It is obvious that this word contains two elements, the element *auto*, and the element *huna*. Now, to be true, *auto* means self, it's a word of Greek origin. And *huna* means the Huna religion from Hawaii, the native religion of the Kahunas. So the term says something like *Do Huna By Yourself*. The meaning of *autohuna* is a special day, one day per year, that the Tsalagi devote to prayer, dance, chant and purification rituals, during which time they practice active forgiveness, and celebrate *friendship*.

This is why commonly, with Western anthropologists, *autohuna* simply is translated as *a friendship day*.

So far so good. But why and how did the Tsalagi know about Huna? This is really an intriguing question. It seems that the original source information cannot be found, so much the more as the tradition of the North-American natives is an *oral tradition*. An eminent expert on this tradition, *Dhyani Ywahoo*, herself a chosen spiritual leader and voice of the elders, writes in her book *Voices of our Ancestors (1988)* that we could only speculate about the ancient connection between the Kahunas in Hawaii and their spiritual brothers,

---

[4] See http://www.autohuna.com

the Tsalagi, in North-America. Fact is that their spiritual rituals are amazingly similar in content and meaning.

This is why I came to use the unusual term *Autohuna* for a site that, while my overall intention as founder of *Ayuda International Foundation* is to help preserving the spiritual and intellectual heritage of all tribal peoples on the globe, is specifically about *Huna*, and only about Huna.[5]

One may wonder why? It's not a personal fancy of mine, to be true. I made this choice, reflecting an *overwhelming tenor* in both anthropological and spiritual literature worldwide that *Huna*, the ancient *Way* of the Kahuna, is one of the most concise, the most elaborated, enlightened and also *the most intelligent and scientific* of all spiritual traditions here on earth.

As a matter of limitation both in available time and in resources, as I have never received any funding nor donations for *Ayuda Foundation*, I must restrict my presentation of the spiritual and scientific heritage that Huna contains on a few literary sources. I also should perhaps add that I never until now had the privilege to meet a Kahuna, nor a Tsalagi native and thus my knowledge must per se be, and come over as, theoretical.

One may pardon me this restriction while it has not done any harm to my enthusiasm and my profound belief and conviction in the wisdom and the importance of native traditions for the survival of humankind in the near coming fu-

---

5 See http://www.ayudainternational.org

ture, and our overall *spiritual progress and evolution* as a human race.

Before I come to give a detailed overview over the *Secret Science*, as Huna is often called, let me shortly outline what sources in our own civilization have to say about the cosmic life energy. While this was not until now a serious issue to lecture about at Western universities, the situation is presently changing dramatically. In fact, the knowledge about the origins of life was relegated in our scientific history to the stoicism of heretic researchers who were each frowned upon, if not discarded out and even outlawed from not only their respective scientific communities, but even society as a whole. I remind only of Wilhelm Reich, who died in jail back in 1957, at a time that many let coincide with the *rational period* of humankind – an assumption that is obviously as wrong as it is idealistic.

Needless to add that Reich was jailed in an American prison not for being a bad American, for having stolen a golden ring, or having slaughtered a human guinea pig.

Reich died for a scientific discovery, for a device called *orgone accumulator* that serves to accumulate the bioenergy and thus can accelerate healing processes in the organism, and that the FDA found, in its eternal system-prone ignorance, to be the invention if not of the devil, but of at least pure charlatanism. And an injunction was issued against this device, by the FDA, containing the firm prohibition to move any of those wooden boxes from Maine, the location of Reich's institute, interstate to any other state of the United States. One day, when Reich was absent from his premises, an assistant of his institute sent one of the accumulators to another state, for research purposes. As a result, Reich was summoned to court, but did not enter an appearance, whereupon a default judgment was rendered, and Reich was arrested for *contempt of court* and jailed, where he died, some time later, from a heart attack.

I mention this here as a reminder for all those (young people) who believe any of humanity's truly great achievements had been brought about with the ease and comfort promised by modern consumer society.

And the second reminder is that the other, and perhaps more traditionally grounded bearers of true knowledge, the natives, were slaughtered in even more cruel ways than a Reich or before him, a Mesmer.[6] By their millions. And on the soil and through the hands of that same *enlightened nation* that likes to think of itself as the *protector spirit* of humankind, the great benefactor of human progress in much the same way as formerly the Christian Church, and that goes around in the world to missionize for *world democracy* and *child protection*.

Beware to keep your mind and your heart on the right side! Most of all, *beware of protectors*, the old ones and the new ones, for they all have an agenda, and most of that agenda is *hidden!*

I do the same and believe me, in now almost twenty years of alternative science publishing, I have *not received one*

---

6 See Pierre F. Walter, *Emonics, Audio Book (2010)* and *Walter's Encyclopedia, Academic Edition (2010)* as well as Pierre F. Walter, *110 Book Reviews (2010).*

single response from any of our great accredited scientists. All my letters and emails to them remained without a response, to this day, and some of them defamed and slandered me not unlike the ways in which Mesmer and Reich were slandered at their times.

In my *Idiot Guide to Emotions (2010)* I have elucidated in some detail the perennial research on the cosmic life energy as it is to be found both in the Eastern and Western science traditions.[7] As the purpose of each book is unique, I cannot avoid repeating myself when presenting the information, and the research, to their different audiences.

While the present guide addresses a lay public and a general science audience, I show the applicability of my scientific approach in the *Idiot Guide to Emotions*, and retrace my research to the Oriental view of the universe as an energy-driven and flowing cosmos.

That guide was largely inspired by my research on *Feng Shui*, the age-old Chinese science of energy-based *vibrational correspondences* in nature and human habitat.[8] It is a rather unusual piece of writing in which I present a complete and practically applicable vocabulary and terminology that could serve as a first step in the creation of a holistic treatment approach for *sexual paraphilias*.[9]

---

[7] See Pierre F. Walter, *The Idiot Guide to Emotions, Awareness Guide (2010)*.

[8] See Pierre F. Walter, *Natural Order, Monograph (2010)*.

[9] Id.

It may be accessible to a rather limited audience of bioenergy researchers and forensic experts, as well as systemically trained psychiatrists. But it also is a selfhelp manual for those who suffer from difficult-to-admit sexual preferences and urges, and will help them to cope constructively with their desire. In this context, then, the guide inscribes as an add-on to my *Emosexcoaching* service, and can serve in some way as a guide manual for my specialized sex coaching.[10]

To my amazement, I found as a result of this research that Wilhelm Reich's discovery of the *orgone* was something not as queer and outlandish as Cartesian thinkers hold it to be, but an event that *had to happen* as a matter of systemic logic. Hence, with enthusiasm and an infinite admiration for Reich's courage and his stout scientific honesty I wrote the second part of this guide, entitled *Reich's Greatest Discoveries*, as well as my more specialized study on Reich's bioenergy treatment approach, *Orgonomy and Schizophrenia*, which represents the third chapter of this guide.

This present guide intends to open the reader's eye to a basically holistic scientific understanding. Let me introduce first what I understand under the term 'secret science' as it could mislead the reader to believe I am talking about just another conspiracy. An eminent expert on the native science tradition, Dhyani Ywahoo, herself a chosen spiritual leader and voice of the elders, writes in *Voices of our Ancestors (1988)* that we can only speculate about the ancient connection between the Kahunas in Hawaii and their spiritual brothers,

---

[10] Id.

the Tsalagi, in North America. Fact is that their spiritual rituals are amazingly similar in content and meaning.

Before I come to give a detailed overview over the *Secret Science*, as Huna is often called, let me shortly outline what sources embedded in our own civilization have to say about the *life force*. While energy science was not until recently a serious issue to write and lecture about, the situation is presently changing dramatically. In fact, the knowledge about the origins of life was relegated in our scientific history to the stoicism of heretic researchers who were each frowned upon, if not discarded out and even outlawed.

I remind the reader only of Wilhelm Reich, who died in jail back in 1957, at a time that many let coincide with the rational period of humanity – an assumption that is obviously as wrong as it is idealistic. Needless to add that Reich was jailed in an American prison not for being a bad American, not for having stolen a golden ring, or for having slaughtered a human guinea pig. Reich died for a scientific discovery, for a device called *orgone accumulator* that serves to accumulate the bioenergy and thus accelerates and facilitates healing processes in the organism, and that the FDA found, in its eternal system-prone ignorance, to be the invention if not of the devil, but of for the least medical charlatanism. And an injunction was issued by the FDA, containing the prohibition to move any of the wooden boxes from Maine, the location of Reich's institute, to any other state of the United States. And one day, when Reich was absent from his premises, an assistant of his institute sent one of the accumulators to another state, for research purposes. As a result,

Reich was summoned to court, but did not enter an appearance, whereupon a default judgment was rendered, and Reich was arrested for contempt of court and put in jail, where he died, some time later, from a heart attack.

I mention this here as a reminder for all younger people who believe any of humanity's greatest achievements had been brought about with the ease and comfort promised by modern consumer society. The second reminder is that the other, and perhaps more *traditionally grounded* bearers of true knowledge, the natives, were slaughtered in even more cruel ways, and by their millions, on the soil and through the hands of the same enlightened nation that likes to think of itself as humanity's first and foremost protector spirit.

Keep your mind and your heart on the right side! Most of all, beware of protectors, the old ones and the new ones, for they all have an agenda, and most of that agenda is hidden!

I do the same and believe me, in now almost twenty years of alternative science publishing, I have not received one single response from one single exemplar among our great accredited scientists, nor have I received any funding. All my letters and emails to members of our glorious science pantheon remained *without a response*, to this day, and some of them defamed and slandered me not unlike the ways in which Mesmer and Reich were slandered.

So you may quietly reflect on that fact, for I think it says more about this great science that our televisions so proudly tell us about, and that pervades our media as the last cry of an enlightened human species! So enlightened in fact that it's at the border not only of insanity, but also of the overkill and worldwide ecological destruction!

In my twenty-seven years of research on the true nature of our emotions, I have seen that all scientists who, over the course of science history, have made out the interactive link or resonance between cell vibration, which I call *emonic vibration*, and health or disease, had a *bioenergetic research approach* which today we would call *systemic*, and they are to be considered the first systems researchers in the history of science! And all of them were able to construe devices or even work without devices to influence and manipulate the cell vibration so as to strengthen immunitary response and fighting pathologies.

The process was particularly elucidative in George Lakhovsky's research in that it was experimentally demonstrated how a simple device, because it triggered *resonance* with the cell's emonic vibration, could actively fight a cancerous tumor in the plant and thus eliminate the cancer.

Hence, we can summarize that all methods and scientific approaches used for assessing, measuring and monitoring vital energies converge in a single well-defined scientific catalogue that is so complete that it can be used as the basis for a new science, a science that integrates the specific knowledge about the cosmic life energy, and that therefore is a functional, systemic and holistic science I suggest to introduce as *The Science of Emonics*.

It has often been said and lectured that the abundant knowledge of native peoples about natural processes was approximative and imaginative, that it was lacking scientific exactitude, and that it was intermingled with religious myths and superstitions. In my *shamanism* research, and as a rather insignificant searcher of truth compared to the great minds and hearts who preceded me in this fascinating gathering of true knowledge, such as Richard Evans Schultes, Ralph Metzner or Michael Harner, I was able to actually invalidate much of the arbitrary or random assumptions that ignorant and, worse, Christian anthropologists notoriously made about native people's intelligence and scientific expertise.[11]

I am going to show in this article that such statements really go astray in view of the sophisticated Huna knowledge tradition that is perhaps the most convincing example for the fact that intelligent people, and an intelligent, lucid culture, will see science and religion as a unique field that radiates toward rationalism, as it radiates toward mysticism, and here I use the term mysticism deliberately to denote the fact that we always will face a black wall in front of the immensity of *complexity and knowledge* that the universe represents.

Here is precisely where religion comes in, that is, at the point where we stand in front of that black wall of ignorance, and must give up our search because, for the moment at least, we can't solve the riddle with the scientific toolset we have developed.

Part of this black wall, as Amit Goswami showed convincingly, is that mechanistic science can't explain subjective phenomena, while we humans are, after all, subjects, and not

---

[11] See Pierre F. Walter, *The Idiot Guide to Consciousness, Awareness Guide (2010)*, Chapter Three, *The Science of Shamanism, Monograph (2010)* as well as *Consciousness and Shamanism, Audio Book (2010)*.

objects in our science. We are the observers, not the observed, hence we are the subjects of our scientific scrutiny or voyage. And as such we are *entangled with our observation*, which means that we cannot honestly claim that we can achieve one hundred percent of objectiveness in any kind of scientific research. That means, a part of the field will always remain black, if we complain about that, it won't change the fact.

Quantum physics has taught us a hard lesson here, it taught us that there is no way out of this maze, other than religion, that is, contemplation of what-is – without judging, without fitting observations in our drawers of past knowledge and tradition. It means we have to remain open for novelty, when we are real scientists!

Down the road, I came to conclude with the late expert on *shamanism*, Terence McKenna, that the true scientists are the natives, and that they preceded our scientific era by centuries, if not millennia, and that thus the humble ones are we and not they. The learners are we, not they. And that's all I say.

THE HERETIC KNOWLEDGE TRADITION

Rediscovering the Real Science

The perennial science I rediscovered over the course of almost twenty years of research was a true *scientia* in the sense the old Greek and Romans used that term; in fact, it is a true *philos sophia*, a science inspired by the love for truth, for wisdom.[12]

Historically however, as I mentioned already, in our Western scientific tradition, scientists and natural healers who acknowledged the existence of the *bioplasmatic energy* were rejected, defamed and persecuted by mainstream science, and some of their books were burnt. And yet this denial of truth that seems to be inherent in Western civilization is not found elsewhere in human history and mythology, as Joseph Campbell said in an interview with Bill Moyers.[13]

---

[12] See Pierre F. Walter, *Natural Order, Monograph (2010)*.

[13] See Pierre F. Walter, *Joseph Campbell and the Lunar Bull, Great Minds Series (2010)*.

All other great civilizations have since millennia acknowledged the fact that life is basically a function of energy and that it is *dynamic and systemic*, and not static and mechanical. And from this principle, that in the Hermetic tradition was metaphorically expressed with 'as it is above, so it is below', it appears in line with functional logic that what is inside the cell will also be enveloping the body.[14]

And in fact, this bioplasmatic energy is both inside the cell and surrounds the physical body like a transparent shell. It folds seven subtle energy layers around the physical or dense body that extent more than just a few inches.[15]

From their erudite energy-based worldview, traditional Chinese and Tibetan medicine as well as *Ayurveda* from India were able to discover in our organism the meridians as the major pipelines of the bioenergetic flow, and could develop the tremendously effective medical science of *acupuncture*.[16] The successes this perennial medical science booked already *thousands of years ago* are still today unheard of in mechanistic, symptom-focused, chemistry-based and palliative Western medicine.

Despite *quantum physics*, which shattered much of the traditional Cartesian and nature-hostile scientific worldview, Western science only reluctantly begins to recognize the fact

---

[14] See Pierre F. Walter, *Natural Order*, Monograph *(2010)*.

[15] See, for example, Shafica Karagulla, *The Chakras (1989)*. See also the important new science sampler *A New Worldview (1996)*, with important contributions by a number of leading-edge scientists.

[16] See Pierre F. Walter, *Natural Order*, Monograph *(2010)*.

that life is holistically coded in *energy patterns* and that no living process can be properly mapped and identified in its functional effectiveness without knowing these patterns and the bioenergy charge they bear. Even avantgarde systems researchers like Rupert Sheldrake or Ervin Laszlo deny what they call the *vitalistic approach* declaring a scientific roof paradigm that integrates the perennial knowledge about the cosmic life energy, against all scientific logic, as *mechanistic*.

Sheldrake goes as far as denying that vitalistic theories are theories at all in the sense of the term; according to him falsifiability, or refutability, or testability lacks with these theories. And here we are left alone, after a cathedral judgment of far-reaching consequences, and which is not followed by any evidence or reference; and such an author is credited with being one of the leading scientists today.

With Ervin Laszlo, it is even worse: he never mentions in his famous and bestselling books any of the authors that I reference in the following paragraph, to come up with an *Integral Theory of Everything* that says all what those authors said over the course of several centuries, but he says it under the header *A-Field*, a term that sounds suspiciously close to Harold Saxton Burr's *L-Field*. And without mentioning Burr with one word!

What Heraclitus, Paracelsus, Goethe, Mesmer, Swedenborg, Freud, Jung, Einstein, Reichenbach, Reich, Lakhovsky, Heisenberg or Bohm had to say about it *goes completely unnoticed.* Such a sworn resistance against knowledge that is perennial is all-too-typical for the arrogant attitude of most Western scientists and their mouthpiece media. I really think that Western science is misguided from the start in that it always had this tendency to discard from its scientific worldview much more than it ever observed and retained. Think only of *Feng Shui*, the heritage of the Druid sages, the fairy worlds, shamanism, plant energies, morphogenetic fields, ectoplasms, telepathy, telekinesis, prophecy, astrology and numerology, or channeling: these branches, and many more, of the great tree of knowledge were cut off, and the result was the rudimentary torso of official Western science, while great minds and brains have *spent lifetimes* researching these fundamental disciplines of *perennial science* all over the world.[17]

Mechanists are unable to *understand nature*, and they can for this very reason not understand a science that, perhaps for the first time in Western science history, is based upon nature and not intellectual assumptions.

My research on the bioenergy was *multi-disciplinary* and cross-cultural both regarding the human bioenergy in the form of emotions, and atmospheric energies that typically are observed by *Feng Shui* or geomancy. The overarching universal creator principle that in religions is attributed to the divine and was given many names, I have called it simply *e*, and this simplicity should exactly reflect its ultimate complexity.

To more aptly describe the creator force or meta-observer of reality, which in the film *What the Bleep Do We Know* was called the *ultimate observer*, I have added-on a second term, *e-force*. While e could be described as the unmanifest

---

[17] Id., Chapter Three, The Twelve Branches of the Tree of Knowledge.

ultimate observer, *e-force* is a more condensed state of manifestation of this creator force; it is the force, the energy, through which *e* is acting upon the surface of consciousness. It is important to realize that *e* and consciousness are *one* in the sense that *e* instantaneously impacts upon and forms consciousness, and that consciousness is aware of *e*. It can be said as well that *e* or *e-force* is contained in consciousness. Now, on the human level, *e-force* has created human emotions, and inhabits emotional energy. Thus, emotional energy is one of the many manifestations of the *e-force*.[18]

Now, as a matter of introduction into what the scientific religion of Huna is about and what kind of knowledge it possesses, let me give you a real-life example of an event that is so far not explainable by conventional science. I cite from *Life after Death (1999)*, by Neville Randall:

> **Neville Randall**
>
> Leslie Flint was said to have a strange and rare gift, the ability to attract the spirits of human beings who had died and moved on to another place of existence, and to provide them with a substance called ectoplasm which they drew from his and his sitters' bodies to fashion a replica of the vocal organs – a voice box or etheric microphone. Through this peculiar contraption which was located about three feet above the medium's head, Woods was told, a spirit transmitted his thoughts. By a process so far unexplainable by science, the desincarnate spirit created vibrations which enabled him to speak to us as using a telephone, in a voice like the one he had on earth.

---

[18] Id., Chapter Three, Toward a Science of Life.

Now, this phenomenon can well be scientifically explained within the methodology of the *Secret Science*. We are observing here what is called by the Kahunas a protruding aka finger, a bioplasmatic substance squeezed out from the body's reservoir of e-force, called *mana*; I call these quanta of vibration *emonic charge*. Actually, this substance acts like a matrix for energetically coded messages and decodes them so that they can be intelligible for people who live in dimensions vibrating at a different frequency from the emitter's.

Through the use of *aka substance*, we can thus construe a translator and transmitter device to help people communicate who live in different vibrational universes. That ectoplasm box is exactly such kind of device.

It not only decodes the emonic vibrations from the other dimension, but also amplifies them so that these vibrations become sound waves intelligible for the human ear.

Of course, conventional science *cannot explain these phenomena*, because it has so far no explanation for the cosmic information field that is the transmitter of these phenomena. It has built a neat house without giving the owner a room for himself. The owner of the house, that I call *e*, is the creator force.

Western science until very recently was a *death science* as it refused to acknowledge the very motor of life, the *life force*. Why this is so has historical reasons. After more than a millennium of life and knowledge denial by the Christian Church, the mechanist scientists built a science that was to *oppose the church*, that had to be a science where god was banned. And for good reason, as all human evolution follows the dynamics of thesis, antithesis and synthesis.

Amit Goswami observes in *The Self-Aware Universe (1995)* that the major weakness of material realism was 'that the philosophy seems to exclude subjective phenomena altogether'. Do we need to wonder, then, that modern science until this day could not inquire into the nature of emotions without facing an abyss?

This can't be different after all since modern science had no idea of the impact of the observer until the severe paradoxes of quantum physics blew up most of Newtonian science and healed that scientific neurosis, in order to reinstate nature, and e, in the house of natural science.

I would have to reference quite a few modern-day authors, who are presently helping to heal the split and who heroically *think different*, in the sense that they help formulating a *truly holistic and coherent science paradigm* of the future. There is definitely a holistic trend now in postmodern science, and a new direction toward integration.

Accordingly, there is a marked change of direction now in Western science that is gradually getting us again in touch with the spiritual dimension that has been discarded out; and from here we are going to step-by-step formulate, probably through a joint effort of many enlightened and spiritually aware scientists a *unified field theory*, or the recognition of *The Field*, as Lynne McTaggart, to mention only her, calls the force or energy that I named e-force.[19]

---

[19] See Lynne McTaggart, *The Field (2002)*.

Carl-Gustav Jung (1875-1961)

Carl-Gustav Jung puts up an astonishing analogy between the Platonic concept of ideas, and the concept of energy in his study *Archetypes of the Collective Unconscious*, saying that basically there is no difference between Plato's *eîdos* concept and Jung's own concept of *psychic energy* that he said was a constituent element in archetypes.[20]

By the way, Jung honestly admits that the term *archetype*, contrary to common belief, is not his invention, but to be found already with Cicero, Pliny, and others and that it also appears in the *Corpus Hermeticum* as a philosophical concept. Looking at the old Greek term for archetype, *to archetypon eîdos*, it becomes clear that an archetype is just one possible form of eîdos. Seen from this perspective, Jung's insight about *ideas containing energy* as archetypes contain psychic energy, is consequent.

The eîdos, Jung explains, are primordial images stored in a supracelestial place as eternal, transcendent forms that the seer could perceive in dreams and visions. From there, Jung pursues:

> **Carl-Gustav Jung**
>
> Or let us take the concept of energy, which is an interpretation of physical events. In earlier times it was the secret fire of the alchemists, or phlogiston, or the heat-force inherent in matter, like the primal warmth of the Stoics, or the Heraclitean ever-living fire, which border on the primitive

---

[20] Carl-Gustav Jung, *Archetypes of the Collective Unconscious (1959)*, 358-407, at 360.

notion of an all-pervading vital force, a power of growth and magic healing that is generally called mana.[21]

We are going to see further down that Jung, in a flash of genius, got a glimpse in what today the quantum physicist Fritjof Capra calls the *Web of Life*.[22] In a couple of sentences Jung draws a synchronistic ellipse from Heraclites over the alchemists to today's still existing tribal cultures that call the cosmic energy *mana*. Jung's insights are the most substantial, as he carefully analyses the nature of what he terms 'psychic energy', and distinguishes it from Freud's *libido* concept and the concept of kinetic energy in atomic physics. In *On The Nature of the Psyche*, Jung writes:

> **Carl-Gustav Jung**
>
> There are indications that psychic processes stand in some sort of energy relation to the physiological substrate. In so far as they are objective events, they can hardly be interpreted as anything but energy processes, or to put it another way: in spite of the non-measurability of psychic processes, the perceptible changes effected by the psyche cannot possibly be understood except as a phenomenon of energy. This places the psychologist in a situation which is highly repugnant to the physicist: The psychologist also talks of energy although he has nothing measurable to manipulate, besides which the concept of energy is a strictly defined mathematical quantity which cannot be applied as such

---

[21] Id., p. 397.

[22] See Fritjof Capra, *The Web of Life (1996/1997)*.

to anything psychic. The formula for kinetic energy, E=mv2/2, contains the factors m (mass) and v (velocity), and these would appear to be incommensurable with the nature of the empirical psyche. If psychology nevertheless insists on employing its own concept of energy for the purpose of expressing the activity (energeia) of the psyche, it is not of course being used as a mathematical formula, but only as its analogy. But note: this analogy is itself an older intuitive idea from which the concept of physical energy originally developed. The latter rests on earlier applications of an energeia not mathematically defined, which can be traced back to the primitive or archaic idea of the 'extraordinarily potent'. This mana concept is not confined to Melanesia, but can also be found in Indonesia and on the east coast of Africa; and it still echoes in the Latin numen and, more faintly, in genius (e.g., genius loci). The use of the term libido in the newer medical psychology has surprising affinities with the primitive mana. This archetypal idea is therefore far from being only primitive, but differs from the physicist's conception of energy by the fact that it is essentially qualitative and not quantitative.[23]

While I question Jung in several points, it is highly interesting, that, as only very few Western psychologists, he has been aware of the perennial concept of a *universal all-pervasive cosmic energy field* that, in accordance with most tribal cultures, he calls *mana*. He never went as far as actually considering this energy field as measurable and not only an archetypal idea; it is useful that he well presents and explains the conceptual problem that is at the basis of his elucidations. How-

---

[23] Carl-Gustav Jung, On the Nature of the Psyche (1959), 47-122, at 130-132.

ever, what Jung points out regarding the difference between *psychic energy* and *kinetic energy* does not stand a deeper analysis. I am going to show further down that there is no basic difference between psychic energy and kinetic energy, but that their apparent difference only stems from the fact that they are measured in different ways.

Jung, rather closed to this idea, states that psychic energy could not be measured, could not be quantified, other than by feeling:

Carl-Gustav Jung

In psychology the exact measurement of quantities is replaced by an approximate determination of intensities, for which purpose, in strictest contrast to physics, we enlist the function of feeling (valuation). The latter takes the place, in psychology, of concrete measurement in physics. The psychic intensities and their graduated differences point to quantitative processes which are inaccessible to direct observation and measurement. While psychological data are essentially qualitative, they also have a sort of latent physical energy, since psychic phenomena exhibit a certain qualitative aspect. Could these quantities be measured the psyche would be bound to appear as having motion in space, something to which the energy formula would be applicable. Therefore, since mass and energy are of the same nature, mass and velocity would be adequate concepts for characterizing the psyche so far as it has any observable effects in space: in other words, it must have an aspect under which it would appear as mass in motion.[24]

---

[24] It., p. 132.

It seems Jung wanted to anticipate *any possible criticism* from the side of his opponents, those who, following a mechanistic paradigm in psychology, would deny the idea of psychic energy as a true dynamic force. And for justifying the energy nature of the psyche, Jung makes an awkward comparison with physics in thinking about the possibility of measurement of the two energies in question: psychic energy on one hand, and kinetic energy, on the other.

First, Jung does not appear to see that *physics itself is mechanistic* when it boasts about the total measurability in physics that already at Jung's lifetime was no more really existing. In fact, only by applying a strictly Newtonian, and thus mechanistic, standard, in physics, we can say that 'all is measurable'. Within the world of subatomic physics or quantum physics, such a science paradigm would however produce wrong or no results. This is so exactly because on that basic, fundamental level, not all is measurable or cognizable, and a large part of the phenomenology is based upon *probability* only.

Jung's problem here, it seems, is his own mechanistic view of psychic energy. First of all, he starts from the premise that 'psychic' and 'kinetic' energy are two different kinds of energy. I would rather take the opposite position and ask, right as the first question in this debate: *Why should we here assume two different kinds of energy?* To me, it makes much more sense in cases of doubt to start from the general paradigm that all *in life is one*, except we can prove it is not. When all is one in nature, we logically have to start from the idea that we deal with the same energy, that however may manifest in different ways. This is namely the crux that Jung has here in his reasoning. He tries to find a common denominator for both energies, something like a *unifying concept*, but then concludes that if psychic energy is like kinetic energy, then the psyche must be something that is in motion, as a mass in motion.

I think we can safely assume that *the psyche is in constant motion*, but that this kind of motion is not one in space, but *one in time*, a constant change and development over time. As time and space, as Einstein's relativity theory clearly says, are intertwined, so must be the two energies, if at all we assume two different energies and not, from the start, one and the same energy manifesting in different ways. Jung concludes:

> **Carl-Gustav Jung**
> If one is unwilling to postulate a pre-established harmony of physical and psychic events, then they can only be in a state of interaction. But the latter hypothesis requires a psyche that touches matter at some point, and, conversely, a matter with a latent psyche, a postulate not so very far removed from certain formulations of modern physics (Eddington, Jeans, and others). In this connection I would remind the reader of the existence of parapsychic phenomena whose reality value can only be appreciated by those who have had occasion to satisfy themselves by personal observation.[25]

These last sentences in Jung's reasoning on psychic energy are stunning in that Jung found a way out of the crux in which he seemed to be caught at the start. Basically, he says

---

[25] Id.

that it would be admissible to advocate both starting points: the admission of a unifying worldview that he, strongly formed by Platonic thought, expresses by the *perennial concept of a state of ideal harmony* of all-that-is, and that he describes as 'pre-established harmony of physical and psychic events', or its contrary.

For the latter presumption, he concludes that also in this case, a kind of synergistic interaction of physical and psychic events, and their respective energies, could not be denied. And to back his statement up, he reminds the reader of parapsychology, psychic research, a discipline that, as we know today, Jung was diligently studying, while at his time, it was less respectable for a psychologist to do psychic research than it is today.

In fact, having done psychic research for more than two decades, I have been stunned over and over again by the fact that basically what we observe in parapsychology are *energy phenomena*, and to a much lesser extent physical, material or touchable events. This was already so in spiritism research, the scientific predecessor of modern parapsychology, and an eminent expert on the matter, Emanuel Swedenborg, was asking the same question as Jung and he answered it by pointing to the bioplasmatic energy that produces, for example, an *ectoplasm* and called it spirit energy.

There is a continuity in bioenergy research in so far as all researchers speak of a unifying energy concept, instead of splitting the cosmic energy into psychic energy, on one hand, and kinetic energy, on the other.

Let me briefly report here, for this purpose, the astonishingly similar explanations of Paracelsus, Swedenborg, Mesmer, Freud, Reichenbach, Reich, and Lakhovsky. It will become obvious that their research corroborates what the secret science knows since millennia.

Paracelsus (1493-1541)

Philippus Aureolus Theophrastus Bombast von Hohenheim, a famous wandering scholar and natural healer from the Germanic part of Switzerland, publishing under the pen *Paracelsus*, was one of the greatest exponents of the perennial pre-Cartesian holistic science, and at the same time a phenomenally successful natural healer and alchemist. He used to call bioenergy *vis vitalis* and identified this energy in all plants.

Paracelsus was the first to recognize that the life force manifests in plants in a way such as to form specific patterns, like a unique identity code assigned to each of them. From this knowledge that is to be found in much the same way in Chinese plant medicine, he lectured that certain plants are collateral for healing and certain others not.

He proposed to take only the essence from these plants, as this was later done by Samuel Hahnemann and Edward Bach in homeopathy, by the use of a distillation process.[26] The healing tinctures he created possessed the distinctive characteristic of being highly effective, condensed and potent agents through their harmonious melting of the various

---

[26] See Pierre F. Walter, *Alternative Medicine and Wellness Techniques, Scholarly Article (2010)*.

plant energies into a higher form of *unison vibration,* which we have to imagine as some sort of composite vibrational code.

The same what Paracelsus did in the West, Chinese sages did in the East, as they found, millennia before his birth, after testing over generations, that no one single plant can reach a healing potency that a set of collateral plants can effect, when they are distinctly distilled into a super-vibrational tincture.[27]

Swedenborg (1688-1772)

Emanuel Swedenborg, known for his research on *spiritism,* called the subtle bioplasmatic energy *spirit energy.* Because of his specific interest in the afterworld, Swedenborg examined the behavior of the bioenergy in ectoplasms and drew his conclusions on the basis of these findings. As a result, Swedenborg lacked the comparative insights that the other researchers could reference, especially those elaborated by Paracelsus and Carl Reichenbach regarding the bioenergetic vibration of plants.

Swedenborg's concept however is well affirming that the cosmic energy is something like a unified concept, contrary to Jung's split definition that acknowledged it only in its dualistic consistence as psychic energy, on one hand, and kinetic energy, on the other.

---

[27] See, for example, Richard Gerber, *A Practical Guide to Vibrational Medicine (2001)* and Donna Eden, *Energy Medicine (1999).* See also Pierre F. Walter, *Richard Gerber and Vibrational Medicine, Book Review (2010)* and *Donna Eden and Energy Medicine, Book Reviews (2010).*

Further, as Swedenborg elaborated an *entire cosmology*, and thus a spiritual explanation of the spirit energy, he ultimately related the life energy to God – as a manifestation of the divine, in much the same way as I have done this in my *Emonics* vocabulary that says that *e-force* is inhabited by *e*, the ultimate creator force.

Mesmer (1734-1815)

Franz-Anton Mesmer was a German (Swabian) physician who, interestingly enough, wrote his doctoral dissertation on the influence of planetary energies upon the human body. His main focus was upon the *Moon* and lunar energy in its influence on various bodily functions such as sleep rhythms, secretion and healing processes. Contrary to Paracelsus' focus upon plants, Mesmer's scientific and medical focus was upon human beings only. He did not consider plants, and as his terminology suggests, saw humans on the same level, energetically speaking, as mammals.

Mesmer, as after him Freud, Charcot or Bleuler, got to his insights through the tedious study of *hysteria* and *female hysterics*. At his time, hysteria, probably because of the widespread cultural sex repression, was a rather *common emotional disease* to be found with middle and upper class women who suffered patriarchal and sex-denying upbringing and who in addition were living in a condition that did not allow them to abreact their sexual energy.

Mesmer's and subsequently Freud's etiology of hysteria was thus sexual, but Mesmer, in good alignment with the morality code of his time, did not touch the sexual question; he simply focused on the vital energy flow and observed that it is somehow influenced by magnetic currents.

He came up with the expression *animal magnetism* for describing the cosmic energy, for the simple reason to distinguish this variant of magnetic force from those which were referred to, at that time, as *mineral magnetism, cosmic magnetism* and *planetary magnetism*. He chose the word 'animal', and not human, because it goes back to *animus*. In Latin, animus means what is 'animated' with life, what breath, what thus belongs to the animate realm. What Mesmer discovered was thus the bioplasmatic energy that was known long before him.

Mesmer first observed healing currents being emitted by huge and strong magnets that he placed between himself and the patient, and later observed, to his astonishment, that the same healing effects occurred also without the magnets. This discovery made him conclude that ultimately it was his own body electrics, his own bioplasmatic vibration that had that curing effect upon his hysteric patients. To conclude, Mesmer can be said to have discovered the subtle energy that before him Paracelsus called *vis vitalis* and Swedenborg *spirit energy* and gave it the somewhat fancy name *animal magnetism*. Except the divergence in terminology, these scientists observed and reported basically the same natural phenomena.

Reichenbach (1788-1869)

Baron Carl Ludwig Freiherr von Reichenbach, a German noble who was a recognized chemist, metallurgist, naturalist and philosopher and member of the prestigious Prussian *Academy of Sciences*, known for his discoveries of kerosene, paraffin and phenol, spent the last part of his life observing the vibrational emanations and bioenergetic code in plants. He spoke of *Od* or *Odic force*, a life principle which he said permeates all living things.

Reichenbach was *by no means a mystic*, but throughout his life a natural scientist. His conclusions were based on the *controlled observation* of natural processes in plants and in humans, and the interactions between plants and humans. For example, when observing a plant in a darkened room in the cellar of his castle that he had isolated against telluric vibrations, he observed, after having accustomed his eyes to the dark for about two hours, a blue-green shadowy egg-formed substance around the plant.

After having been certain about his own accurate perception and the repeatability of the experiment, he invited other scientists and lay persons to join him in his observations, and all the other persons, who were carefully selected in terms of mental clarity and sanity, corroborated his observation.

On the basis of his astounding discoveries, Reichenbach set out to heal sick people with the Odic force construing various devices for this purpose. He became very popular as he, as a very rich industrial, went to the poor to heal their suffering family members. His research clearly corroborates an important part of the spiritual microcosm of the native Kahunas in Hawaii and the corresponding cosmology of the Cherokee natives in North America who almost exclusively use plant-contained bioenergy in their approach to heal disease.

Reich (1897-1957)

Dr. Wilhelm Reich was a physician and psychoanalyst, and later orgone researcher, from Austria. Reich was a respected analyst for much of his life, focusing on *character structure*, rather than on individual neurotic symptoms. Reich was in many ways far ahead of his time in promoting healthy adolescent sexuality, the availability of contraceptives and abortion, and the importance of economic independence for women. He is best known for his studies on the link between human sexuality and emotions, the importance of what he called orgastic potency, and for what he said was the discovery of a form of energy that permeated the atmosphere and all living matter, which he called *orgone*. He built boxes called *orgone accumulators*, in which patients could sit, and which were intended to accumulate the bioenergy.

Reich corroborated through his *orgone research* what holistic researchers before him already had observed: that life is coded in patterns of an invisible subtle bioplasmatic energy that is not to be confounded with bioelectricity, and that is somehow related to the creator principle.[28]

Lakhovsky (1869-1942)

Georges Lakhovsky was a Russian electrical engineer and scientist who emigrated to France before World War I. In 1929, Lakhovsky published his book *Le Secret de la Vie* in Paris, translated in English as *The Secret of Life*. Lakhovsky discovered that all living cells possess attributes normally associated with electronic circuits. Observing that the oscillation of high frequency sine waves when sustained by a small, steady supply of energy would bring about resonance, Lakhovsky must be credited with the original discovery of today we know as *cell resonance*.

Lakhovsky found that not only do all living cells produce and radiate *oscillations of very high frequencies*, but that they also receive and respond to oscillations imposed upon them by external sources. In fact, Lakhovsky attributed the source of radiation to cosmic rays that constantly bombard the earth. From these insights, Lakhovsky construed devices for healing with high frequency waves, that today we know as *Radionics*.[29] Lakhovsky found that when outside sources of oscillations are resonating in synch with the energy code of the cell, the

---

[28] See, for example, Wilhelm Reich, *The Function of the Orgasm (1942)*, *The Cancer Biopathy (1973)*, *The Mass Psychology of Fascism (1933)*, *Selected Writings (1973)*, *Children of the Future (1950)*, *Record of a Friendship (1981)*, Myron Sharaf, *Fury on Earth (1983)*.

[29] See, for example, David K. Tansley, *Chakras, Rays and Radionics (1984)*.

growth of the cell would become stronger, while when frequencies differed, this would weaken the vitality of the cell. From this initial observation, Lakhovsky further found that the cells of pathogenic organisms produce different frequencies than normal, healthy cells.

Lakhovsky specifically observed that if he could increase the amplitude, but not the frequency, of the oscillations of healthy cells, this increase would dampen the oscillations produced by disease causing cells, thus bringing about their decline. However, when he rose the amplitude of the disease-causing cells, their oscillations would gain the upper hand and as a result the test person or plant would become weaker and illness increase. As a result, Lakhovsky viewed the progression of disease as essentially a battle between resonant oscillations of host cells versus oscillations emanating from pathogenic organisms.

He initially proved his theory using plants. In December, 1924, he inoculated a set of ten geranium plants with a plant cancer that produced tumors. After thirty days, tumors had developed in all of the plants, upon which Lakhovsky *took one of the ten infected plants* and simply fashioned a heavy copper wire in a one loop, open-ended coil about thirty centimeter (12") in diameter around the center of the plant and held it in place. The copper coil was found to collect and concentrate energy from extremely high frequency cosmic rays. The diameter of the copper loop determined which range of frequencies would be captured. Lakhovsky found that the thirty centimeter loop captured frequencies that fell within the resonant frequency range of the plant's cells.

This captured energy thus reinforced the resonant oscillations naturally produced by the nucleus of the geranium's cells. This allowed the plant to overwhelm the oscillations of the cancer cells and thereby destroy the cancer. The tumors fell off in less than three weeks and by two months, the plant was thriving. All of the other cancer-inoculated plants, those that were not receiving the copper coil, died within thirty days.

Lakhovsky then fashioned *loops of copper wire for humans* that could be worn around the waist, neck, elbows, wrists, knees, or ankles and found that over time relief of painful symptoms was obtained.

These simple coils, worn continuously around certain parts of the body, would invigorate the vibrational strength of cells and increased the immune response which in turn took care of the offending pathogens.

Upon which Lakhovsky construed a device that produced a *broad range of high frequency pulsed signals* that radiate energy to the patient via two round resonators: one resonator acting as a transmitter and the other as a receiver.

The machine generated a wide spectrum of high frequencies coupled with *static high voltage charges applied to the resonators*. These high voltages cause a corona discharge around the perimeter of the outside resonator ring that Lakhovsky called *effluvia*.

The patient would sit on a wooden stool in between the two resonators and was exposed to these discharges for about fifteen minutes. The frequency waves sped up the recovery process by stimulating the resonance of healthy cells in the patient and in doing so, increased the immune response to the disease-causing organisms.

Burr (1889-1973)

Harold Saxton Burr was E. K. Hunt Professor Emeritus, Anatomy, at Yale University School of Medicine. Burr found that all living things are molded and controlled by *electrodynamic fields* and demonstrated to measure them using standard voltmeters. He named them fields of life or simply the *L-field*. Beginning in the 1930s with his seminal work at Yale, Burr was able to verify his initial hypothesis of subtle energy fields that govern the human body. Burr set up a series of experiments that showed that all living organisms are surrounded and encompassed by their own energy fields. He showed that changes in the electrical potential of the *L-field* would lead to changes in the health of the organism. By leaving some trees on the Yale campus hooked up to his *L-field* detectors for decades, he was able to demonstrate that changes in environmental electromagnetic fields such as the phases of the moon, sunspot activity, and thunderstorms, substantially affected the *L-field*.

He found he could detect a specific field of energy in a frog's egg, and that the nervous system would later develop precisely within that field, suggesting that the *L-field* was the organizing matrix for the body.

In his work with humans, he was able to chart and predict the ovulation cycles of women, to locate internal scar tissue, and to diagnose potential physical ailments, all through the reading of the individual's *L-field*. Student and colleague Leonard Ravitz carried Burr's work forward.

Ravitz focused especially on the human dimension, beginning with a demonstration of the effects of the lunar cycle on the *L-field*, reaching a peak of activity at the full moon. Through work with hypnotic subjects, he demonstrated that changes in the *L-field* directly relate to changes in a person's mental and emotional states. Ravitz came to the conclusion that emotions can be equated with energy. Most intriguingly, Ravitz showed that the L-field as a whole disappears before physical death.

While Burr expressed himself in a rather *misleading terminology*, speaking of 'electricity' when he connoted the life force, and of 'electromagnetic fields' when it was about the human energy field, but most of the literature on energy and vibrational medicine cite Burr as one of their pioneers.

In fact, Masaru Emoto says in his book *The Secret Life of Water (2005)* about Burr that he 'laid much of the basic foundation for the science of hado'.[30]

---

[30] Masaru Emoto, *The Secret Life of Water (2005)*, p. 139.

THE SECRET SCIENCE

The Huna Knowledge of the Cosmic Energy Field

Among native populations, there is a tradition called 'universal doctrine' by Joseph Campbell and that is consistent with observing and recognizing the existence of a universal energy. In *The Hero With a 1000 Faces (1973/1999)*, Campbell writes:

> **Joseph Campbell**
> Briefly formulated, the universal doctrine teaches that all the visible structures of the world – all things and beings – are the effects of a ubiquitous power out of which they rise, which supports and fills them during the period of their manifestation, and back into which they must ultimately dissolve. This is the power known to science as energy, to the Melanesians as mana, to the Sioux Indians as wakonda, the Hindus as shakti, and the Christians as the power of God.

Its manifestation in the psyche is termed, by the psychoanalysts, libido.[31]

Among natives, the Kahunas possess perhaps the most *systemic understanding* of the bioenergetic coding of life, and it is surely from them that the Sioux and the Cherokee of North America adopted it. The religion of the Kahunas, as Max Long, an American psychologist, found in his lifelong research on Huna, considered the knowledge about *mana, the cosmic energy,* as a secret science. Long observes:

Max Long

> It was a virgin field because, in spite of startling evidence of the powers of the kahunas (the priests and magic-workers of olden times), anthropologists had tossed their works and beliefs into the discard as 'superstition'. The Christian missionaries, arriving in 1820, disapproved of miracles performed by natives, and bent every effort toward eradicating kahuna beliefs.[32]

Max Long found that these natives excel by their *specific ability to understand human consciousness* and the fact that consciousness and cosmic life energy are basically one. Contrary to our knowledge that in this field was mainly conceptualized by early psychoanalysis, the Kahunas regard the unconscious, that they call *unihipili*, as a spirit force, and not as a trash container. And they ascribe to this force a certain inde-

---

[31] Joseph Campbell, *The Hero with a Thousand Faces (1973/1999)*, pp. 257-258.

[32] Max Long, *The Secret Science at Work (1995)*, p. 1.

pendence of will and intention. By its own will, this force, that they call the *lower self*, may stop collaborating with the other inner selves. Further, the Kahunas are teaching that it is the lower self that manufactures and handles the organism's *mana*, its vital energy reservoir. Long named the current or flow of this energy 'auric charge'.

In fact, the idea that energy and consciousness are linked in some way is very old and it is some sort of intuitive knowledge. As Joseph Campbell observes toward Bill Moyers in *The Power of Myth (1988):*

> **Joseph Campbell**
> I have a feeling that consciousness and energy are the same thing somehow. *Where you really see life energy, there's consciousness.*[33]

Mana, the Kahunas believe, is the vital force, the *life force*, and this force is being observed and attributed concise characteristics. This force is said, for example, to be the constituent of all of the activities of the three selves. Max Long observes that the Kahuna priests taught that the lower self creates mana 'automatically ... from food eaten and air breathed'.[34]

Max Long reports he found through slow and patient effort that the Kahunas' belief in the three selves describes each of these selves as an entity that dwells 'in three invisible or shadowy bodies, one for each self'. This shadowy body

---

[33] Joseph Campbell, *The Power of Myth (1988)*, p. 18.

[34] Max Long, *The Secret Science at Work (1953/1995)*, p. 10.

was named *aka body* by the Kahunas, while esoteric sciences, as Long rightly remarks, use to call them 'etheric doubles'.[35]

Long saw that the Kahunas used a handy metaphor for describing the *mana* force; they associated it with water as a liquid substance that represents the juice of life; from this basic idea, the Kahunas extrapolated the metaphor of the human being as a tree or plant, 'the roots being the low self, the trunk and branches the middle self, and the leaves the high self'. While the sap circulating through roots, branches and leaves vividly illustrated the nature of the mana force.[36]

The Essenes, the first Christians, interestingly represented the same or a very similar imagery regarding the vital force. It was for this reason, as Edmond Bordeaux-Szekely found, that they had given so much importance to the water purification ritual. In fact, the Essenes spoke of a *Goddess of the Water*, a vital force that was inhabiting water and that can purify us through the use of daily cold showers taken in free nature and with water taken directly from a source such as a mountain stream well known to contain highly pure water.[37]

Now, the amazing research on water and vibrations by the Japanese natural healer Masaru Emoto fully confirms these findings with new and surprising evidence.[38] Emoto

---

[35] Id.

[36] Id., p. 11.

[37] Dr. Edmond Bordeaux-Szekely, *Gospel of the Essenes (1988)*.

[38] See Pierre F. Walter, *Natural Order, Monograph (2010)* and *Do You Love Einstein?, Monograph (2010)*.

found the enormous implications of vibration by looking at the vibrational code of water that he calls *hado*. In the Japanese spiritual tradition, hado is indeed considered as a vibrational code that, similar to *ki*, the life energy, has healing properties and transformative powers. Literally translated, hado means wave motion or vibration. Once we become aware of hado's flow, Dr. Emoto showed, it can spark positive changes in our physical space and emotional wellbeing. What Emoto teaches can be called *hado awareness* or vibrational awareness, as an integral part of our general acute awareness of how we influence our environment through thoughts and emotions. The point of departure is thus to recognize and acknowledge that in every thought and emotion, a specific *vibration* manifests.

Masaru Emoto's research was greatly promoted through the metaphysical documentary film *What the Bleep Do We Know*, but existed long before the great public got to learn about it through he movie. These findings have shown that the crystalline structure of water can be influenced by feelings, intentions, sounds. It is important to note that these findings confirm a basic insight of *Feng Shui*, which says since thousands of years that only flowing water contains *ch'i*, while stagnant water contains deadly energy or *sha*. Feng Shui, therefore, considers only flowing water to contain the positive life force, while stagnant water is deemed to contain a rather harmful and retrograde variant.[39]

---

[39] *Sha* is probably equivalent to Wilhelm Reich's notion of *Deadly Orgone (DOR)*.

The next amazing discovery that Emoto came about was the fact that water has a memory – a memory far longer than our transient lifetimes. And third, that we can learn from water, by allowing it to resonate within us. Dr. Emoto writes in *The Secret Life of Water (2005)* that hado has essentially four characteristics. They are *frequency, resonance, similarity* and *flow*.[40] And this is equally valid for our emotions! They have a frequency, they show patterns of resonance, they follow the laws of similarity and they are in constant flow. Emotions have a frequency because they vibrate. They are vibrations. And their frequency is unique. Emoto writes:

Masaru Emoto
Frequency can be modeled as waves, a fact easily supported by quantum mechanics. All matter is frequency as well as particles. What this means is that rather than considering something a living organism or a mineral, something we can touch or something we can see, everything is vibrating, and vibrating at a unique and individual frequency.[41]

Regarding the low self, the Kahunas believe that its aka body can slide *into and out of the physical body* and that it impregnates every cell and tissue of the body and brain. The aka body is seen as a mold of every cell or tissue or fluid. It is in this etheric body, the aka body of the low self, that the Kahunas situate the emotions. They believe that love, hate

---

[40] Masaru Emoto, *The Secret Life of Water (2005)*, pp. 33-35.

[41] Id., p. 30.

and fear all come from the low self as emotions. By contrast, they teach that the major job of the middle self is to learn to control the low self and prevent it from running off with the man.

In this context, it is especially of interest how the Kahunas explain the nature of prayer, namely as the low self contacting the high self by means of the *aka cord*, which it activates, and along which it sends a supply of *mana* used by the high self in answering the prayer. In the spiritual microcosm that our human organism represents for the Kahunas, sensory impressions are believed to be received through the organs of the five senses, and presented to the middle self for explanation. The middle self is depicted as the reasoning self, what we today use to call our rational mind, while the low self's task is thought to be one of perceiving and recording. It is said that the low self makes a tiny mold of the aka substance of its shadowy body, something like recording sound on a tape while all sounds, sights, thoughts or words are believed to come in patterns called 'time trains', which are functional units containing many single impressions joined together. More precisely, the Kahunas symbolize these as *clusters of small round things* such as grapes or berries. Ordinarily, these microscopic clusters of invisible substance are thought to carry *mana* in that part of the aka body of the low self which impregnates or identifies itself with the brain. At the time of death, the Kahunas teach, the low self in its aka body leaves the body, and in doing so takes with it the memories.

The Kahunas' scientific spirituality is so refined that they even set out to explain phenomena such as *hypnosis*. They actually believe that hypnosis is a way to produce thought forms of ideas that are implanted in the aka body of the one willing to accept the suggestion. The same is true for *time travel* that the Kahunas explain as the fact that the entire aka body of the low self projects itself into a distance, *connection with the physical body being maintained by a cord of aka substance.*[42]

Finally, the perhaps most sophisticated scientific achievement of the Kahunas is their explanation of *memory*. They in fact relate memory to thought forms and explain these as energy patterns within the low self. A number of related impressions is believed to form a *cluster of thoughtforms*, and such clusters are thought to record and contain the memories of complete events. By the same token, those memory clusters are believed to reside in the aka body of the low self rather than in the physical brain tissues.

Max Long observes that medical discoveries have demonstrated that the aka of the brain, during life and consciousness, interblends with corresponding parts of the physical brain, and that openings cut in the skull to bare the outer layer of the brain in the region above and behind the ears, can be touched with a needle carrying a mild electric current, and, without injury to the patient, can cause him to remember and even live over in vivid detail events of his past life.[43]

---

[42] Id., p. 38.

[43] Id., p. 50.

Long also reports about a device for measuring the mana current called 'aurameter' and that preceded by several years the discovery of the human, animal and plant auras by *Kirlian Photography*. Long found that the exact dimension of the aka body or aura of any living being can be made out with this device. He observed that 'normally, the aka protrudes only a few inches from the body except at the shoulder blades and over the genitals, at which points the aura extends farther'.[44] He also writes that tests using the aurameter showed that the spirits of the dead survive and live in their *aka bodies* all around us.[45] Here is what he explains:

> **Max Long**
>
> Mr. Mark Probert of San Diego, a well-known medium, has a number of spirits who come to speak through him when he is in a trance condition. On this occasion, he went into the customary trance and a spirit spoke through his lips, carrying on a lively conversation and showing much interest in the Aurameter which was being tested. He readily agreed to stand beside the medium while Mr. Cameron tried to locate his aka body and trace its outline. He found it at once, and outlined it with as much ease as if it had belonged to a living man.[46]

Regarding the size of the *aka body*, Long noted a peculiarity that he said the Kahunas are well aware of, that the

---

[44] Id.

[45] Id., p. 57.

[46] Id.

visualized aka form often seems to have grown or contracted very much, when found. The Kahunas, Long reports, believe that the *aka body* could be made large so that it protrudes greatly, or so small that it retreats inside the body, and that thought forms have the same quality.[47] In so far, Long observes, the Kahunas teach that the middle self plays its part by deciding what each event means and what its relation to other events may be – or, as they say, rationalizing it:

> **Max Long**
>
> The memory cluster of thought forms, once it has been given its rational meaning and significance by the middle self, is stored by the low self in the aka body.[48]

With the same amazing clarity and simplicity, the Kahunas explain telepathy, believing that '… the mana flows along the aka cord between two people who are in telepathic communication'.[49] Long pursues:

> **Max Long**
>
> The invisible aka threads or cords may be likened roughly to telegraph wires over which messages can be sent. They carry mana much as wires carry electricity. Just as the telegraph wires carry symbol messages to the receiving end, the aka threads can and do carry – on the flow of mana

---

[47] Id., p. 58.

[48] Id., p. 59.

[49] Id., p. 61.

running through them – clusters of microscopic thought-forms.[50]

The most interesting in Long's research about the Kahunas' spiritual microcosm is the nature of the *mana force*. He said right away that it certainly is not electricity of the electromagnetic type, and that it acts more like direct current of the type generated through chemical action.

> **Max Long**
>
> However, it is characterized by the fact that it seems to be a living force when aka body or aka cord substance serves as a storage place for it, or as a conducting wire or rod or cord. It has another characteristic in that it seems to find in the aka substance a perfect conductor.[51]

This is Long's summary report of the Kahunas' concise teaching of telepathy:

> **Max Long**
>
> In telepathy we have proof that the aka thread is a perfect or living substitute for a wire, and that the mana flows as easily over a connecting thread half way around the world as across a room. The popular theory that telepathic sending is similar to the sending of high frequency radio waves through the air, as in a broadcast, has been proven a fallacy. The radio waves fade and weaken inversely as the square of the distance traveled, and with a power plant as small as the

---

[50] Id.

[51] Id., p. 62.

human low self, a broadcast of this type would hardly be able to reach farther than a few feet.[52]

And it was 'with nothing but their aka bodies and mana taken from the living to fill them', that spirits, according to Long, during séances, use up all the mana in a single sudden effort with the result that the living can be lifted into the air, tables or heavy pianos lifted, or even entire houses shaken as by an earthquake.[53]

In addition, Long writes, spirits could strike through *aka lasers* that 'would render the warrior struck temporarily unconscious, much as the mesmerist in Hollywood, by projecting a surcharge along the line of his vision – undoubtedly with a projected finger of aka-mana, and that could send a man sprawling to lie unconscious on the floor'.[54]

What is especially noteworthy for the present guide is that the Kahunas knew that the *life force* is effectively manipulated by the impact of consciousness, which exactly confirms Joseph Campbell's intuition.

As a result of their basically scientific worldview, the Kahunas have *no moralistic roof structure* as all our great dominator civilizations and they know only one sin: that of hurting another, and this also only in the case that when this hurt is

---

[52] Id.

[53] Id., p. 79.

[54] Id., pp. 79-80.

done when being fully aware of it and yet doing it against better knowing.[55]

Yet the Kahunas' secret science is by far not the only source of this knowledge, while it's well outstanding in its detailed and scientific investigation and presentation. Walter Y. Evans-Wentz, in his research on the fairy faith in Celtic countries, came across this knowledge as well. Wentz observes in his book *The Fairy Faith in Celtic Countries (1911)* that an Irish mystic and erudite on the fairy faith regarded fairy paths or *fairy passes*, the locations where fairies habitually appear, as magnetic arteries through which circulates the earth's magnetism. In addition, he reports that the water fairies are said to be kept alive 'by something akin to electrical fluids'.[56]

Dr. Ong Hean-Tatt, a bioenergy researcher from Malaysia, wrote a concise study about the scientific basis of *Feng Shui*, the five thousand years old energy science of the Chinese and concluded from a wealth of observations and discoveries that this science deals with the cosmic energy using about the same precision and objectivity as Newtonian physics regarding gravity.[57]

As I have shown in my review of this important book, Dr. Ong establishes amazing parallels between *Feng Shui* and the perennial knowledge about the telluric force known as

---

[55] Id., p. 91.

[56] Walter Y. Evans-Wentz, *The Fairy Faith in Celtic Countries (1911/2002)*, p. 33, note 1.

[57] Ong Hean-Tatt, *Amazing Scientific Basis of Feng Shui (1977)*.

geomancy, which has a long-standing tradition in both the East and the West. The factual evidence produced by the author that relates in detail various UFO sightings and reports from reputed sources is dumbfounding and seems to prove the fact that these phenomena feed upon earth energies or telluric energies emanating from underground water.

He also found that important religious cult sites, such as *Stonehenge*, were built exactly on the intersection of telluric lines; not astonishingly so, it's exactly around these sites that most of spirit, angels, ghost and UFO sightings actually occur, and for the very reason that these places are flooded with cosmic energy and therefore allow other dimensions to connect with ours through *energetic cross-section and vibrational resonance*.

Further, Dr. Ong examines the *bird migration phenomenon* and finds that it corroborates the evidence forwarded for the existence of a *telluric world grid*, the fact is that the birds more or less follow those lines and that the energy that emanates from them serves the birds as a navigation help.

In his conversations with Bill Moyers, Joseph Campbell speculates that all gods in all religions are ultimately but energy manifestations:

> **Joseph Campbell**
>
> [T]he gods are rather manifestations and purveyors of an energy that is finally impersonal. They are not its source. The god is the vehicle of its energy. And the force or quality of the energy that is involved or represented determines the character and function of the god. There are gods of violence, there are gods of compassion,

there are gods that unite the two worlds of the unseen and the seen, and there are gods that are simply the protectors of kings or nations in their war campaigns. These are all personifications of the energies in play. But the ultimate source of the energies remains a mystery.[58]

---

[58] Joseph Campbell, *The Power of Myth (1988)*, p. 259.

GLOSSARY

Contextual Glossary

Complexity

Complexity is a major characteristics of living systems. Generally in all *flow patterns*, complexity and simplicity are complementary opposites. This is so not only in natural phenomena, but also in ontology and in human psychology. This duality has been recognized by the ancient Mesoamerican natives. There is an intricate relationship between <u>consciousness</u> and complexity; complexity is a function of the vital energy and information flow; when energy and information flow freely in the organism, complexity tends to be high, while it's reduced when the flow of energy and information is blocked or obstructed.

As a matter of evolution, life and particularly human life tends to increase in complexity over time. However, historically and socially, every time when shifts of consciousness occur that bring about a marked increase in complexity, a counter-reaction sets in that typically, and propagandistically, denies complexity and begins to threaten, persecute and socially discard individuals, and especially scientists who research on areas of human complexity that are not yet fully understood and that are therefore surrounded with taboo, confusion and fear.

Social historians such as *Jacob Burckhardt (1818-1897)*, and psychologists have found the deeper reasons of fascism in a deep-rooted fear of change, and an almost paranoid fright in facing complexity, and especially sexual complexity. You find in child abuse generally a denial of complexity, as the child for the abuser is an object for self-satisfaction and emotional security, and thus a thing to manipulate. Generally speaking, behind protection is often hidden a deep longing for domination and manipulation!

Humans are complex not only generally, but also sexually. In the ancient Indian tradition of sexual Tantra, this wisdom has been orally transmitted from generation to generation and it is even today still part of native religions such as Huna. In the west, early European Orientalists originally reviled Tantra as a subversive, antisocial, licentious and immoral force that had corrupted classical Hinduism. However, many today consider Tantra positively, as a celebration of social equity, sexuality and the body.

Consciousness

Consciousness basically consists of three major elements:

- Perception
- Information Processing
- Energy

The most important part of my scientific observation of consciousness is that it contains *energy*, the information field, or human energy field, so that energy is a constituent part of it, next to perception and information processing.

In Western scientific history, the energy part of consciousness has been consistently blinded out from scrutiny and occulted, to a point that in Western cultures, there is a huge knowledge gap about the human energy field as a result of this cultural prohibition of the 'tree of knowledge'. Consequently, my consciousness research is focused upon bringing in the missing links so as to arrive at a *unified field of integrative perception* and thus a coherent model of consciousness.

Bibliography

David Bohm
Wholeness and the Implicate Order (2002)
Thought as a System (1994)

Gregory Bateson
Steps to an Ecology of Mind (2000)

Fritjof Capra
The Turning Point (1987)
The Web of Life (1997)
The Hidden Connections (2002)

Amit Goswami
The Self-Aware Universe (1995)

Stanislav Grof
Beyond the Brain (1985)
The Holotropic Mind (1993)
The Three Levels of Human Consciousness (1993)

Hameroff, Newberg, Woolf, Bierman, Ramachandran, Chalmers
Consciousness
20 Scientists Interviewed
Director: Gregory Alsbury
5 DVD Box Set (2003)

Dean Radin
The Conscious Universe (1997)

Lynne McTaggart
The Field (2002)

Michael Talbot
The Holographic Universe (1992)

Direct Perception

Direct Perception is the primary mode of learning that nature applies in evolution. Direct perception is the mode the human brain uses to receive and store information in its capacity as a passively organizing system. The child learns his or her first language through direct perception, the picking-up of whole patterns, using the integrative and associative mode of the right brain. Obedience and imitation are not the appropriate means to develop the human potential; therefore civilization can only function on an outside or superficial level, but not as a motor of integrating man into a truly functional power unit that is operating on all levels at once.

The mainstream educational system has put this natural intelligent and holistic learning mode upside down in forcing children to learn with their left brain hemisphere only, cutting off the necessary mode of synthesis provided by the right brain hemisphere. This is the single major reason why the modern educational system, while it is very costly, is totally ineffective, and brings about people who are alienated from their own inner source, out of touch as they are with their innermost human potential. This also is the reason for the astonishing lack of creativity in the corporate world, that already coach and corporate trainer Edward de Bono deplored in his books.

E-Force

See Emonics

Emonics

Definition

Emonics is a science vocabulary I have created. The name is an abbreviation of *Emotional Identity Code Science*. Purpose of the vocabulary is to facilitate scientific investigation of human emotions, especially in the form of emotional identity in the etiology of *sexual paraphilias*. The paraphilia research that I conducted over more than twenty years has given me conclusive evidence that every human being possesses a unique emotional identity code, something like a vibrational ID number, that works like a cosmic identifier and sets us apart as absolutely unique beings. This is valid not only for humans but this vibrational pattern is unique also for animals, for plants and even for inanimate matter such as rocks.

Emonics research allocates to human emotions a quality that is much different from traditional research on emotions.

While traditional psychology has to some extent admitted the cognitive nature of emotions, it relates emotions to thought and perception only and locates them in the brain. Emonics, *in accordance with a number of perennial science traditions and cutting-edge research on the human energy field* shows that emotions are *located in the human aura* and possess an inherent quality of *flow*.

Emonics research clearly shows that thought and emotions are vibrations that flow through our etheric or luminous body. In this sense, also animals and plants do have emotions, which was something completely discarded or overlooked by traditional psychology. Hence, Emonics research can be said to transcend psychology and to some extent unify biology, psychology, parapsychology and physics into something like a *unified field theory of emotions* that holistically inquires into the nature of emotional identity.

Emonics Vocabulary

E and E-Force

E, the creator identity, is the functional complement of *consciousness*. Consciousness is a function of e in that e and consciousness are a functional whole. E manifests on this planet as *e-force*. Shifts in consciousness bring about shifts in e-force, which in turn trigger altered states of consciousness. Superconsciousness is a state of e-force at its peak level.

Emonic and Demonic

Disintegration of emotions occurs through repression and denial. The results are violence, regression and sadism – which are obstacles to evolution.

Emonic Charge

The biogenic positive charge accumulated in living organisms leads, typically, to discharge in the form of ecstatic convulsions or sexual orgasm and is part of the inherent self-regulatory system of the cell plasma.

Emonic Awareness

The conscious perception of our *emonic flow* includes the lucid awareness of our emotional predilection and sexual attraction in every given moment or situation. For example, nurses should be conscious of their organism's emonic flow regarding patients they are working with; educators need to develop emonic awareness regarding their natural pedoemotions projected upon some of the children they are working with. In fact, I am using the term emonic awareness synonymously with the term emotional awareness.

Emonic Current or Emonic Flow

The bioenergetic current flows through the organism, from the cell plasma to the periphery and into the luminous body and again back from the luminous body to the cell, depending on the polarity of the current. When it is positive, it is expansive and flows from the cell to the periphery (joy); when it is negative, it retreats from the periphery back into the cell (fear). Emonic flow, in popular language, may be expressed as <u>emotional flow</u>, and I do indeed use both terms synonymously.

These flow principles inherent in the nature of the bioenergy are also to be applied in the etiology of <u>sadism</u>. In the natural sexual streaming of the bioenergy, that Wilhelm Reich described as 'hot, melting streamings', the energy during orgasm explodes from the cell toward the luminous body. In <u>sadism</u>, however, because of the muscular armor in the pelvis region and other parts of the body, the energy cannot freely flow outwardly and therefore is repelled back with the result that instead of relaxing joy and expansive feelings, what is felt after orgasm is depression, anxiety, and fatigue. These latter symptoms then, can also be used as signals in diagnosing sadism.

As a result of these insights, it is possible to actually heal sadism by getting the emonic current again to flow naturally through the entire organism. This can be done through muscular *relaxation* or through consciousness work, using *Creative Prayer*, or *Life Authoring*, or else a combination of these with methods practiced by *Alternative Medicine*, such as body work, massage, *Qigong, Tai Chi Chuan, Reiki*, or *Yoga*.[59]

---

[59] See Pierre F. Walter, *The Life Authoring Manual (2010)*, *The Idiot Guide to Soul Power (2010)* and *Alternative Medicine and Wellness Techniques (2011)*.

Emonic Integrity

Children and babies naturally, when they are swinging in their continuum balance, are within the realm of emonic integrity.

Emonic Sanity

Emonic sanity is manifest when emotional energy is integrated, which is the natural condition in the living organism. This can also be called *emotional balance*. Emonic sanity if further characterized by high complexity and high emotional and erotic intelligence. Integration occurs ideally on three levels:

- Multisensorial (Spirituality)
- Extrasensorial (Parapsychology)
- Sensorial (Eroticism, Sexuality)

Emonic Sanity in Relationships With Children

Emonic sanity with children is a task of every parent and every educator; the task consists in caring for preserving the natural continuum balance of the child. This means in practice to observe a principle of *sacred non-interference* in the child's continuum, to restrain from inflicting educational violence on the child, to respect the child's privacy, to actively foster the child's social and erotic life, which means to abstain from controlling the child's relationships/friendships with peers and other adults, to give the child real opportunities for love and sexual relations outside of the family, to restrain from emotional and sexual incest with the child, and to help the child accept their body and their emotions through loving dialogue about all life situations, without taboo.

Emonic Setup

Emonic setup means our natural bioenergetic setup from birth, the free flow of the vital energies in our organism, the healthy vibration of aura and bioplasma, the natural cycle of charge and discharge in our sexual embrace, during the whole of the life cycle from conception to death. Our emonic setup is by nature harmonious and self-regulated, and it favors equitable relationships, love and natural sharing of emotions, joy, and goodness.

It becomes distorted through early interference with the natural energy pattern in form of educational violence and abuse, and the obstruction of the emonic flow through the educational moralistic prohibition of expressing emotions and sexual wishes through truthful dialogue.

Emonic Vibration

Emonic vibration is the bioenergetic flow and unique vibrational code that is inherent in every living organism, and without which life would cease and death would occur. Emonic vibration is thus an immediate characteristic of life.

Emotional Flow

Emotional Flow is a synonymous term to *emonic flow*, a notion I have developed in the context of my research on sexual paraphilias, and it describes the natural flow condition of our emotions, when no distortion has taken place in the bioplasmatic setup, through a life-alienating *moralistic* education, and/or the suffering of emotional abuse in the form of an ongoing co-dependence with one parent, or with both parents, in childhood and/or adolescence.

Quantum Physics

Quantum mechanics (QM, or quantum theory) is a physical science dealing with the behaviour of matter and energy on the scale of atoms and subatomic particles / waves. QM also forms the basis for the contemporary understanding of how very large objects such as stars and galaxies, and cosmological events such as the Big Bang, can be analyzed and explained. Quantum mechanics is the foundation of several related disciplines including condensed matter physics, quantum chemistry, molecular biology, particle physics, and electronics. The term 'quantum mechanics' was first coined by Max Born in 1924. The acceptance by the general physics community of quantum mechanics is due to its accurate prediction of the physical behaviour of systems, including systems where Newtonian mechanics fails. Even general relativity is limited in ways quantum mechanics is not, for describing systems at the atomic scale or smaller, or at very low or very high energies, or at the lowest temperatures.

Through a century of experimentation and applied science, quantum mechanical theory has proven to be very successful and practical. (Wikipedia)

Definition

Quantum Physics or quantum mechanics is a fundamental branch of theoretical physics with wide applications in experimental physics that replaces classical mechanics and classical electromagnetism at the atomic and subatomic levels.

It is the underlying mathematical framework of many fields of physics and chemistry, including condensed matter physics, atomic physics, molecular physics, computational chemistry, quantum chemistry, particle physics, and nuclear physics. Along with general relativity, quantum mechanics is one of the pillars of modern physics.

The Uncertainty Principle

The once certain basic assumptions about life, that were the pillars of Cartesian science, were replaced by uncertainty. It was through Werner Heisenberg that this often-quoted *uncertainty principle* was established in physics, and notoriously much to the exasperation of Albert Einstein who reportedly objected 'God does not play dice!'

Nonlocality

Another basic discovery of quantum physics is *nonlocality*. Nonlocality means that effects be triggered by element A in element B without element A and element B having any form of physical connection. They can in fact be light years away from each other. Nonlocality, then, is not bound to relativity, and effects therefore are not a function of the speed of the light nor any higher velocity; in other words, they are instantaneous. The term used for nonlocal effects is *entanglement* or *quantum en-*

tanglement. An alternative explanation was given by Rupert Sheldrake who explains nonlocal effects by *morphic resonance*.[60]

Quantum physics has been my light of hope since more than a decade. It has given me right in all my childhood intuitions, and in my violent reject of the arrogant ignorance that was sold by our science teachers as the only truth existing in the cosmos.

Extrapolating this personal experience I have come to tell children to actively resist being raped by educational systems, that until today have never understood the human brain, and therefore do not know how to properly educate a child for acquiring a stable basis of viable knowledge that later can be used in real life. Einstein knew why he dropped out of all schools and universities, and Picasso as well; there is a long list to be made of other geniuses who did the same.

Quantum physics was for me *something like a heavenly punishment* for all those idiot science teachers with their ignorant assumptions about life and their eternal neurosis, who bury our lively and moving children in graves called schools, thereby transforming them into living mummies.

Quantum physics with all its apparent paradoxes has for me a direct connection with the Divine in that it forces the conditioned human spirit to relax and sit back in front of the true mystery of life, and let God. Quantum physics also has something of the Hindu God *Shiva, the Destroyer*, in that it has destroyed the hubris of many make-belief scientists, leading all of us toward renewed humility in front of the immense intelligence that is

---

[60] See, for example, Fritjof Capra, *The Tao of Physics (1975/2000)*, Deepak Chopra, *Life After Death (2006)*, Russell DiCarlo, *A New Worldview (1996)*, Amit Goswami, *The Self-Aware Universe (1995)*, Ervin Laszlo, *Science and the Akashic Field (2004)*, Lynne McTaggart, *The Field (2002)*, Rupert Sheldrake, *A New Science (1995)*, Michael Talbot, *The Holographic Universe (1992)*, Russell Targ, *Miracles of Mind (1999)*, Vidette Todaro-Franceschi, *The Enigma of Energy (1999)*.

behind Creation. I think all these paradoxes that quantum physics so abundantly serves us are the result of the moron intellectual assumptions that traditional modern science was putting up as 'eternal principles of living' but that in fact have little to do with the object of observation, that is, *nature*. And that is why they eventually are turning out as what they are: *projections upon nature, not the outcome of observation of nature.*

Ervin Laszlo explains in Science and the Akashic Field:

> **Ervin Laszlo**
>
> In the course of the twentieth century, quantum physics – the physics of the ultrasmall domain of physical reality – became strange beyond imagination. The discoveries show that the smallest identifiable units of matter, force, and light are actually made up of energy, but not a continuous flow of energy: they always come in distinct packets known as quanta. These energy packets are not material, although they can have matterlike properties such as mass, gravitation, and inertia. They seem like objects, but they are not ordinary, commonsense objects: they are both corpuscles and waves. When one of their properties is measured, the others become unavailable to measurement and observation. And they are instantly and nonenergetically 'entangled' with each other no matter how far apart they may be.[61]

The Uncertainty Principle

Today a school teacher cannot that easily indoctrinate children with pseudo-scientific assumptions that are but the outflow of his moralism and his reductionist life paradigm, as this was the case still during my childhood and school times. This is so be-

---

[61] Ervin Laszlo, *Science and the Akashic Field (2004)*, p. 31.

cause the certainty has been replaced by uncertainty. And it was perhaps through <u>Werner Heisenberg</u> and his oft-quoted *uncertainty principle* that quantum physics was established as a science. Vidette Todaro writes in her elucidating study *The Enigma of Energy (1991)*:

> **Vidette Todaro-Franceschi**
> It was but one small leap from the uncertainty principle and the dual wave-particle character of matter to physicist Niels Bohr's theory of complementarity. He proposed that on the quantum level nothing can be divided into discrete parts. Everything is related; everything complements everything else. Since it is impossible to completely predict outcomes on a quantum level, we are forced to look at the whole.
> Furthermore, Bohr illuminated the philosophical issues entwining complementarity in the quantum physics world with psychic experience and the study of living organisms in general. The simple fact is that everything is inseparably connected.[62]

This has done us a lot of good because this is more than just physics. It's life. What impact this can have upon healing, Dr. Villoldo recognized it in his bestselling book *Shaman, Healer, Sage (2000)*. He writes:

> **Alberto Villoldo**
> Physicist Werner Heisenberg developed a key principle of quantum mechanics: that one could determine either the velocity or the position of an electron accurately, but not both. The Heisenberg uncertainty principle states that the act of observing an event influences its outcome or destiny. Heisen-

---

[62] Vidette Todaro-Franceschi, *The Enigma of Energy (1991)*, p. 37.

> berg's discovery seems to indicate that the ability to change the physical world through the exercise of vision is very limited once energy has manifested into form. The time to change the world is before form has emerged from the formless, before energy has manifested into matter. Thus many of the healing practices developed by shamans heal conditions before they manifest in the body, before old imprints in the Luminous Energy Field have organized matter into illness or misfortune.[63]

Life is basically uncertain and to sell life to children as something certain is a lie, an immense and punishable lie, no way! To make up a worldview of certainty in front of children means to render them *stiff and neurotic,* and this is exactly what our school system has done in the past, and in many countries still today does. But in some of the most advanced countries, this new science today in some way benefits our school system and renders teachers humbler and more uncertain in their attitude toward children. And this uncertainty is truly a blessing!

For, to paraphrase, Carl Gustav Jung, it's those who had ideal parents and teachers that fill the waiting rooms of psychiatrists all over the world! Those ideal parents and teachers namely have one main characteristic: they appear in front of their children as *certain about everything,* life, love, God, sin, and punishment.

They appear to be certain about themselves. They appear to be certain about the world. They appear to be certain about destiny. They appear to be certain about science and about discipline. And that is why they are simply *useless as parents and teachers.* And many of them would benefit from some lessons in quantum physics, or a good and well-paid trip to the mental hospital!

---

[63] Alberto Villoldo, *Shaman, Healer, Sage (2000),* p. 131.

One more basic discovery of quantum physics is *nonlocality*. Nonlocality means that effects be triggered by element A in element B without element A and element B having any kind of physical connection. They can in fact be light years away from each other. Nonlocality, then, is not restricted to relativity theory, and effects therefore are not a function of the speed of the light nor any higher velocity; hence, they are instantaneous.

How can that be? This is one of the most explosive topics in quantum physics, while it's only one of many paradoxes that it has brought to implode in the body of modern physics.

As I am not a physicist, I will quote Dean Radin, who writes in his book *Entangled Minds (2006)*:

Dean Radin

At a level of reality deeper than the ordinary senses can grasp, our brains and minds are in intimate communion with the universe. It's as though we lived in a gigantic bowl of clear jello. Every wiggle – every movement, event, and thought – within that medium is felt throughout the entire bowl. Except that this / particular form of jello is a rather peculiar medium, in that it's not localized in the usual way, nor is it squishy like ordinary Jell-O. It extends beyond the bonds of ordinary spacetime, and it's not even a substance in the usual sense of that word.

Because of this nonlocal Jell-O in which we are embedded, we can get glimpses of information about other people's minds, distant objects, or the future or past. We get this not through the ordinary senses and not because signals from those other minds and objects travel to our brain. But because at some level of our mind/brain is already coexistent with other people's minds, distant objects, and everything else. To navigate through this space, we use attention and intention. From this perspective, psychic experiences are reframed not as mysterious

> 'powers of the mind' but as momentary glimpses of the entangled fabric of reality.
> Particles that are quantum entangled do not imply that signals pass between them. Entanglement means that separated systems are correlated.[64]

I am not one who is certain, and I will not as a non-physicist lecture about such a difficult matter as is quantum physics. What I am primarily interested in is the *effect that quantum physics has on the transformation of current patriarchal society*. This transformation is real, and that, to repeat it, is my light of hope.

What the Bleep Do We Know?!

You can learn also a lot about quantum physics from the film *What the Bleep Do We Know!?* and by additionally following up with reading the main publications of the scientists who were featured in the film. Let me quote some, and some authors who were unfortunately not featured in the film, and which I equally reviewed. Get your information first-hand, and avoid books and web sites from non-physicists who pretend to being able to explain quantum physics. In most cases you will get a distorted picture.

On the other hand, do not be misled to think that you won't understand a bit when you read first-hand literature, written by famous quantum physicists. Fact is that most quantum physicists today follow the outstanding example that Fritjof Capra has set many years ago, who was perhaps the first widely published physicist gifted with the talent to explain complex matters in easy and understandable language.

---

[64] Dean Radin, *Entangled Minds (2006)*, pp. 263-264.

Capra's example has impressed many of his professional colleagues and today you can say that every quantum physicist makes it an honor for himself or herself to lecture about the matter in a way that a greater public can easily follow and understand.

As quantum physics really is a prototypical example of complexity in action, the literature I recommend is not limited to the explanation of quantum mechanics in the strict sense. To give an example of a book that I have not reviewed here because it's quite impossible to review it, as it's so complex, and so 'all over the place'. It's Jeffrey Satinover's book *The Quantum Brain (2001)*, a book that very artfully weaves quantum mechanics, brain research and psychiatry into one epic tale of grandiose dimensions that offers a broad outlook on the possibilities of a future humanity. The book is a good example how poetic a scientist may become when he thoroughly gets involved with quantum mechanics. And the interesting thing about Satinover is that he is a trained psychiatrist and only in a 2nd life study cycle has become a quantum physicist. This is extraordinary in itself!

Another book, which I have reviewed, is Deepak Chopra's *Life After Death (2006)*, a book that is apparently only about life after death, but it's also about quantum physics, and without our discovery of quantum physics, this book could not have been written, or it could well have been written, but most people simply would not understand and accept the book as it is. This is a fact and has nothing to do with Chopra's popularity. Quantum physics has opened certain pathways in our brain that before were lying dormant.

Sadism

Sadism is a blockage of the natural emotional flow through a predominantly moralistic or puritanical education, often accompanied by physical punishment, which leads to a repression of the natural streaming of the hot and melting sexual energy and as a result, to *demonic emotions*, and violence, because the naturally deep sexual discharge becomes shallow or even is inhibited.

As a result, the naturally hot and tender sexual feelings are disintegrated and distorted into a *compulsion for sex* targeting at strong explosive sexual discharge, as a matter of abreacting an urge, instead of embracing a mate. Sexual discharge in fact temporarily alleviates the fear armor but tends to entangle the person, who is more or less unconscious of the affliction, long-term in sexual aggression, assault and generally a bullying, racketing or abasing behavior, that degrades and dehumanizes the mate to a passive dummy.

Sadism was badly understood before Reich's in-depth research on the sexual orgasm revealed that the natural sexual drive is by no means aggressive or compulsive, but controlled by empathy and love for the sexual mate. Only in sadism, which is a distortion of the natural emotional and sexual setup, this empathy tends to be overridden by an overwhelming longing for egocentric, and power-ridden, satisfaction virtually on the back, and to the detriment, of the sexual mate. This is why long-term sexual sadism leads to a corruption of the personality, as the pattern for abuse then is laid also in a general manner, and the person tends to take advantage of others in the form of a habitual behavior structure, and thus becomes what is called an abuser.

But for this to happen, the pattern *must have been ingrained for long*, and the person must never have gained conscious awareness about it. This is rather the extreme case, as often people become aware of their sadistic needs and begin to become suspicious of the obvious violence of their sexual behavior, and then begin to look for a way out, and may seek out a minister, physician, psychiatrist or psychotherapist for advice and consultation.

Breaking the sadism pattern is greatly facilitated by being around babies and small children, and generally, when men are actively involved in taking care of things, of children, of trees, of gardens and flowers, or of cooking and cleaning the house.

Hence, the need for involving males in early child care. All these tasks are getting men in touch with their *yin* side, or anima, thereby helping them to overcome the macho or hero spirit that is negatively conducive to building the abuse pattern as a long-term affliction and personality trait. For we have to see that sadism is not only an individual problem, but also a societal concern.

Our Western culture is largely sadistic and this sadism can be shown and demonstrated with many examples from the historian's or the psychohistorian's toolbox. Thus, sadism is a direct outflow and consequence of centuries if not millennia of moralism as a sort of emotional plague that has distorted our emosexual behavior structure. Our value system is deeply freedom and touch hostile and this value system was built because our deep emotions are *out of touch with our natural emosexual base structure*. This value system is against nature because it favors violence and shuns natural sexual tenderness and respectful nonviolent embrace among generations, as a prolongation of necessary and health-fostering touch among all members of society.

Shamanism

Definition

Shamanism is a way of apprehending reality, a set of insightful techniques, rituals and patterns, as well as a natural and organic lifestyle centered not at dominating nature or cosmos, but at *participating* in and *understanding* of nature and cosmos. The most important to find out about shamanism is its use of *entheogens*. These are plants that contain psychoactive compounds, such as DMT, and others, and that, when taken at appropriate doses, produce a consciousness-altering effect upon our psyche and perception.

The *shaman* typically is the one who stands out because of his unique capability to explore, and travel into different realities and levels of *consciousness*.

Entheogens

While there are methods to alter consciousness without plants, using esoteric breathing techniques, body postures or ecstatic dance, prayer, fasting and other techniques, researchers agree that from a point of view of effectiveness there is a large gap between those latter techniques, and the use of entheogenic drugs. Entheogens are several hundred percent more effective than non-plant based methods.

Several researchers have seriously tackled the question why this is so, and one of the most persisting on this specific point was Terence McKenna. In his book *The Archaic Revival (1992)*, he affirms that entheogenic plants contain the very essential genetic code, the basic information about the evolution of life on earth, and that for this reason their ingestion, or at least the ingestion of the psychoactive compounds they contain, leads to an immediate opening of consciousness, which was something much broader and much more intelligent to experience than mere colorful visions. In fact, McKenna's visionary and illuminating books would never have had such a powerful impact on the consciousness change of Western society if they only had talked about some nice hallucinogenic visions. Anthropologists or generally researchers who try to understand the unique phenomenon of shamanism and reduce the entheogenic experience to a mere social game, a distraction or pleasure, or a search for some kind of artificial nirvana, are deeply misled.

It is therefore not surprising that most anthropologists, and especially those of them who really do not understand shamanistic cultures, tend to employ expressions such as 'hallucinogens', 'narcotic drugs', 'narcotics' or 'psychedelics'. Apart from the fact that these plants are *not* narcotics, because a narcotic drug, such as for example *opium*, renders somnolent but does not alter consciousness, the important thing to know is that entheogens are not understood, in shamanistic cultures, as leisure drugs, but really are considered as assets of the *religious and numinous experience*. That is why the only expression that comes close to the shamanistic mindset is the term *entheogens*, that is facilitators for getting in touch with the inner god.

It has equally been found that entheogens, apart from their helping us to reach the inner mind, also dissolve habits such as alcoholism, and generally help in a process of social deconditioning. Entheogens help us to see behind the veil of the normative behavior code in any given society as they show us different behavior options. We can thus ingest these substances as a sort of 'social medicine'; we will then be able to recognize the patterns of normative behavior we are caught in and that obstruct our creativity and self-realization.

People who are socially oppressed, racial, ethnic, religious or sexual minorities, may want to inquire into the possible dissolution of rigid behavioral rules and oppressive normative standards in society. They may want to look for the ultimately most intelligent catalyzer that exists to see all the options reality offers and, as a result, might want to engage in a consciousness-opening voyage.

Another important observation regards mental health. It has often been wrongly stated that indigenous shamanic populations were psychotic or pre-psychotic and it comes to mind that usually so-called pedophiles, in our society, once convicted and subjected to 'psychiatric expertise', are labeled in exactly the same way. When we remember the times of communism in Russia or read books by Aleksandr Solzhenitsyn and others, we learn that under that totalitarian regime the same murderous psychiatry with exactly the same vocabulary had been used to eliminate intellectuals who were treated as system enemies because they defended human rights and democratic values.

All this may not surprise any informed individual. The effective mechanisms to defend a given societal 'standard behavior' paradigm are all founded not upon natural pleasure-seeking behavior but upon adaptive perversity. Hence, the necessity for us to look beyond the fence of behavior patterns and inquire into realms that seem apart from it but aren't. The human soul expresses its originality always in paradoxes and sometimes in extreme behavior. The attempt to 'classify' human behavior into rigid 'standards for all' is in itself an ideology – idiotology. The more a given society puts up general standards, the more it is alienated from life and its creative roots and the more it is *subject to decay and perversion.*[65]

---

[65] See, for example, Mircea Eliade, *Shamanism (1964)*, Piers Vitebsky, *Shamanism (2001)*, Ralph Metzner (Ed.), *Ayahuasca (1999)*, Michael Harner, *Ways of the Shaman (1990)*, Jeremy Narby, *The Cosmic Serpent (1999)*, Richard E. Schultes et al., *Plants of the Gods (2002)*, Terence McKenna, *The Invisible Landscape (1994)*, *True Hallucinations (1998)*, *The Archaic Revival (1992)*, *Food of the Gods (1993)*, Robert Forte (Ed.), *Entheogens and the Future of Religion (2000)*, Luis Eduardo Luna, Pablo Amaringo, *Ayahuasca Visions (1999)*, Adam Gottlieb, *Peyote and Other Psychoactive Cacti (1997)*, Aldous Huxley, *The Doors of Perception and Heaven and Hell (1954)*, Rick Strassman, *DMT: The Spirit Molecule (2001)*, Josep M. Fericla, *Al trasluz de la Ayahuasca (2002)*.

BIBLIOGRAPHY

General Bibliography

A

Abrams, Jeremiah (Ed.)
Reclaiming the Inner Child
New York: Tarcher/Putnam, 1990

Die Befreiung des Inneren Kindes
Die Wiederentdeckung unserer ursprünglichen kreativen Persönlichkeit und ihre zentrale Bedeutung für unser Erwachsenwerden
München: Scherz Verlag, 1993

Adrienne, Carol
The Numerology Kit
New American Library, 1988

Agni Yoga Society
COEUR : Signes de l'Agni Yoga
Toulon: Sté Edipub, 1985
Publication originale date de 1932

Albrecht, Karl
The Only Thing That Matters
New York: Harper & Row, 1993

Alston, John P. / Tucker, Francis
The Myth of Sexual Permissiveness
The Journal of Sex Research, 9/1 (1973)

Appleton, Matthew

A Free Range Childhood
Self-Regulation at Summerhill School
Foundation for Educational Renewal, 2000

Summerhill
Kindern ihre Kindheit zurückgeben
Demokratie und Selbstregulierung in der Erziehung
Hohengehren: Schneider Verlag, 2003

Arcas, Gérald, Dr

Guérir le corps par l'hypnose et l'auto-hypnose
Paris: Sand, 1997

Ariès, Philippe

L'enfant et la famille sous l'Ancien Régime
Paris, Seuil, 1975

Centuries of Childhood
New York: Vintage Books, 1962

Geschichte der Kindheit
Frankfurt/M: DTV, 1998

Arntz, William & Chasse, Betsy

What the Bleep Do We Know
20th Century Fox, 2005 (DVD)

Down The Rabbit Hole Quantum Edition
20th Century Fox, 2006 (3 DVD Set)

Bleep
An der Schnittstelle von Spiritualität und Wissenschaft
Verblüffende Erkenntnisse und Anstösse zum Weiterdenken
Berlin: Vak Verlag, 2007

Arroyo, Stephen

Astrology, Karma & Transformation
The Inner Dimensions of the Birth Chart
Sebastopol, CA: CRSC Publications, 1978

Astrologie, Karma und Transformation
Die Chancen schwieriger Aspekte
Frankfurt/M: Heyne Verlag, 1998

Relationships and Life Cycles
Astrological Patterns of Personal Experience
Sebastopol, CA: CRCS Publications, 1993

Handbuch der Horoskop-Deutung
Berlin: Rowohlt, 1999

Atlee, Tom

The Tao of Democracy
Using Co-Intelligence to Create a World That Works for All
North Charleston, SC: Imprint Books / WorldWorks Press, 2003

B

Bachelard, Gaston

The Poetics of Reverie
Translated by Daniel Russell
Boston: Beacon Press, 1971

Poetik des Raumes
Frankfurt/M: Fischer Verlag, 2001

Bachofen, Johann Jakob

Gesammelte Werke, Band II
Das Mutterrecht
Basel: Benno Schwabe & Co., 1948
Erstveröffentlichung im Jahre 1861

Baggins, David Sadofsky
Drug Hate and the Corruption of American Justice
Santa Barbara: Praeger, 1998

Bagley, Christopher
Child Abusers
Research and Treatment
New York: Universal Publishers, 2003

Balter, Michael
The Goddess and the Bull
Catalhoyuk, An Archaeological Journey
to the Dawn of Civilization
New York: Free Press, 2006

Bandler, Richard
Get the Life You Want
The Secrets to Quick and Lasting Life Change
With Neuro-Linguistic Programming
Deerfield Beach, Fl: HCI, 2008

Barbaree, Howard E. & Marshall, William L. (Eds.)
The Juvenile Sex Offender
Second Edition
New York: Guilford Press, 2008

Barnes, A. James, Dworkin, Terry and Richards Eric L.
Law for Business, 9th Edition
New York: McGraw-Hill, 2006

Barnes, J. (Ed.)
The Complete Works of Aristotle, Vol. 1
Princeton: Princeton University Press, 1971

Barron, Frank X., Montuori, et al. (Eds.)
Creators on Creating
Awakening and Cultivating the Imaginative Mind
(New Consciousness Reader)
New York: P. Tarcher/Putnam, 1997

Bateson, Gregory
Steps to an Ecology of Mind
Chicago: University of Chicago Press, 2000
Originally published in 1972

Bender Lauretta & Blau, Abram
The Reaction of Children to Sexual Relations with Adults
American J. Orthopsychiatry 7 (1937), 500-518

Benkler, Yochai
The Wealth of Networks
How Social Production Transforms Markets and Freedom
New Haven, CT: Yale University Press, 2007

Bennion, Francis
Statutory Interpretation
London: Butterworths, 1984

Bernard, Frits
Paedophilia
A Factual Report
Amsterdam: Enclave, 1985

Pädophilie ohne Grenzen
Theorie, Forschung, Praxis
Frankfurt/M: Foerster Verlag, 1997

Kinderschänder?
Pädophilie, von der Liebe mit Kindern
3. Auflage
Frankfurt/M: Foerster Verlag, 1982

Bertalanffy, Ludwig von
General Systems Theory
Foundations, Development, Applications
New York: George Brazilier Publishing, 1976

Besant, Annie
An Autobiography
New Delhi: Penguin Books, 2005
Originally published in 1893

Karma
4e édition
Paris: Adyar, 1923

Bettelheim, Bruno
A Good Enough Parent
New York: A. Knopf, 1987

The Uses of Enchantment
New York: Vintage Books, 1989

Kinder brauchen Märchen
Frankfurt/M: DTV, 2002

Beutler/Bieber/Pipkorn/Streil
Die Europäische Gemeinschaft
Rechtsordnung und Politik
2. Auflage
Baden-Baden: Nomos, 1982

Block, Peter
Stewardship
Choosing Service Over Self-Interest
San Francisco: Berrett-Koehler, 1996

Blofeld, J.
The Book of Changes
A New Translation of the Ancient Chinese I Ching
New York: E.P. Dutton, 1965

Blum, Ralph H. & Laughan, Susan
The Healing Runes
Tools for the Recovery of Body, Mind, Heart & Soul
New York: St. Martin's Press, 1995

Boadalla, David
Wilhelm Reich, Leben und Werk
Frankfurt/M: Fischer, 1980

Bodin, Jean
On Sovereignty (1576)
Six Books of the Commonwealth
Edited by Professor Julian Franklin
New York: Seven Treasures Publications, 2009

Böhm, Wilfried
Maria Montessori
2. Auflage
Bad Heilbrunn: Julius Klinkhardt, 1991

Bohm, David
Wholeness and the Implicate Order
London: Routledge, 2002

Die implizite Ordnung
Grundlagen eines dynamischen Holismus
München: Goldmann Wilhelm, 1989

Thought as a System
London: Routledge, 1994

Quantum Theory
London: Dover Publications, 1989

La plénitude de l'univers
Paris: Rocher, 1992

La conscience de l'univers
Paris: Rocher, 1992

Boldt, Laurence G.

Zen and the Art of Making a Living
A Practical Guide to Creative Career Design
New York: Penguin Arkana, 1993

How to Find the Work You Love
New York: Penguin Arkana, 1996

Zen Soup
Tasty Morsels of Zen Wisdom From Great Minds East & West
New York: Penguin Arkana, 1997

The Tao of Abundance
Eight Ancient Principles For Abundant Living
New York: Penguin Arkana, 1999

Das Tao der Fülle
Vom Reichtum, der uns glücklich macht
Mittelberg: Joy Verlag, 2001

Bordeaux-Szekely, Edmond

Teaching of the Essenes from Enoch to the Dead Sea Scrolls
Beekman Publishing, 1992

Gospel of the Essenes
The Unknown Books of the Essenes
& Lost Scrolls of the Essene Brotherhood
Beekman Publishing, 1988

Gospel of Peace of Jesus Christ
Beekman Publishing, 1994

Gospel of Peace, 2d Vol.
I B S International Publishers

Das Friedensevangelium der Essener
Saarbrücken: Neue Erde/Lentz, 2002

Évangile essénien de la paix
La vie biogénique
Genève: Éditions Soleil, 1978

Die unbekannten Schriften der Essener
Saarbrücken: Neue Erde/Lentz, 2002

Branden, Nathaniel
How to Raise Your Self-Esteem
New York: Bantam, 1987

Die 6 Säulen des Selbstwertgefühls
Erfolgreich und zufrieden durch ein starkes Selbst
München: Piper Verlag, 2009

Brant & Tisza
The Sexually Misused Child
American J. Orthopsychiatry, 47(1)(1977)

Brassai
Conversations with Picasso
Chicago: University of Chicago Publications, 1999

Brennan, Barbara Ann
Hands of Healing
A Guide to Healing Through the Human Energy Field
New York: Bantam, 1988

Brongersma, Edward

Aggression against Pedophiles
7 International Journal of Law & Psychiatry 82 (1984)

Loving Boys
Amsterdam, New York: GAP, 1987

Das verfemte Geschlecht
Berlin: Lichtenberg Verlag, 1970

Bruce, Alexandra

Beyond the Bleep
The Definite Unauthorized Guide to 'What the Bleep Do we Know!?'
New York: Disinformation, 2005

Bullough & Bullough (Eds.)

Human Sexuality
An Encyclopedia
New York: Garland Publishing, 1994

Sin, Sickness and Sanity
A History of Sexual Attitudes
New York: New American Library, 1977

Burgess, Ann Wolbert

Child Pornography and Sex Rings
New York: Lexington Books, 1984

Burwick, Frederick

The Damnation of Newton
Goethe's Color Theory and Romantic Perception
New York: Walter de Gruyter, 1986

Butler-Bowden, Tom

50 Success Classics
Winning Wisdom for Work & Life From 50 Landmark Books
London: Nicholas Brealey Publishing, 2004

50 Klassiker des Erfolgs
Die wichtigsten Werke von Kenneth Blanchard, Warren Buffet, Andrew Carnegie, Stephen R. Covey, Spencer Johnson, Benjamin Franklin, Napoleon Hill, Nelson Mandela, Anthony Robbins, Brian Tracy, Sun Tsu, Jack Welch und vielen anderen
Frankfurt/M: MVG Verlag, 2005

50 Lebenshilfe Klassiker
Frankfurt/M: MVG Verlag, 2004

50 Klassiker der Psychologie
Die wichtigsten Werke von Alfred Adler, Sigmund Freud, Daniel Goleman, Karen Horney, William James, C.G. Jung, Jean Piaget, Viktor Frankl, Howard Gardner, Alfred Kinsey, Abraham Maslow, Iwan Pawlow, Stanley Milgram, Martin Seligman und vielen anderen
Frankfurt/M: MVG Verlag, 2004

50 Klassiker der Spiritualität
Die wichtigsten Werke von Augustinus, Khalil Gibran, Mahatma Ghandi, Dag Hammarskjölkd, Hermann Hesse, C. G. Jung, Eckhart Tolle, J. Krishnamurti, Thich Nhat Hanh, Mutter Teresa, Dan Millman und vielen anderen
Frankfurt/M: MVG Verlag, 2006

Buxton, Richard

The Complete World of Greek Mythology
London: Thames & Hudson, 2007

#

Cain, Chelsea & Moon Unit Zappa

Wild Child
New York: Seal Press (Feminist Publishing), 1999

Calderone & Ramey
Talking With Your Child About Sex
New York: Random House, 1982

Campbell, Herbert James
The Pleasure Areas
London: Eyre Methuen Ltd., 1973

Der Irrtum mit der Seele
München: Scherz Verlag, 1973

Les principes du plaisir
Paris: Stock, 1974

Campbell, Jacqueline C.
Assessing Dangerousness
Violence by Sexual Offenders, Batterers and Child Abusers
New York: Sage Publications, 2004

Campbell, Joseph
The Hero With A Thousand Faces
Princeton: Princeton University Press, 1973
(Bollingen Series XVII)
London: Orion Books, 1999

Der Heros in Tausend Gestalten
München: Insel Verlag, 2009

Occidental Mythology
Princeton: Princeton University Press, 1973
(Bollingen Series XVII)
New York: Penguin Arkana, 1991

The Masks of God
Oriental Mythology
New York: Penguin Arkana, 1992
Originally published 1962

Mythologie des Ostens
Die Masken Gottes Bd. 2
Basel: Sphinx Verlag, 1996

The Power of Myth
With Bill Moyers
ed. by Sue Flowers
New York: Anchor Books, 1988

Die Kraft der Mythen
Düsseldorf: Patmos Verlag, 2007

Cantelon, Philip L. (Ed.)

The American Atom
A Documentary History of Nuclear Policies from the
Discovery of Fission to the Present
With Richard G. Hewlett (Ed.) and Robert C. Williams (Ed.)
Philadelphia, PA: University of Pennsylvania Press, 1992

Capacchione, Lucia

The Power of Your Other Hand
North Hollywood, CA: Newcastle Publishing, 1988

Capra, Bernt Amadeus

Mindwalk
A Film for Passionate Thinkers
Based Upon Fritjof Capra's *The Turning Point*
New York: Triton Pictures, 1990

Capra, Fritjof

The Turning Point
Science, Society And The Rising Culture
New York: Simon & Schuster, 1987
Original Author Copyright, 1982

Wendezeit
Bausteine für ein neues Weltbild
München: Droemer Knaur, 2004

Le temps du changement
Science, société et nouvelle culture
Paris: Rocher, 1994

The Tao of Physics
An Exploration of the Parallels Between Modern
Physics and Eastern Mysticism
New York: Shambhala Publications, 2000
(New Edition) Originally published in 1975

Das Tao der Physik
Die Konvergenz von westlicher Wissenschaft und östlicher Philosophie
Neue und erweiterte Auflage
München: O.W. Barth bei Scherz, 2000
Ursprünglich erschienen 1975 bei Droemersche Verlagsanstalt
in Hamburg

Le tao de la physique
Paris: Sand & Tchou, 1994

The Web of Life
A New Scientific Understanding of Living Systems
New York: Doubleday, 1997
Author Copyright 1996

Lebensnetz
Ein neues Verständnis der lebendigen Welt
München: Scherz Verlag, 1999

The Hidden Connections
Integrating The Biological, Cognitive And Social
Dimensions Of Life Into A Science Of Sustainability
New York: Doubleday, 2002

Verborgene Zusammenhänge
München: Scherz, 2002

Steering Business Toward Sustainability
New York: United Nations University Press, 1995

Uncommon Wisdom
Conversations with Remarkable People
New York: Bantam, 1989

The Science of Leonardo
Inside the Mind of the Great Genius of the Renaissance
New York: Anchor Books, 2008
New York: Bantam Doubleday, 2007 (First Publishing)

Complete List of Publications
http://www.fritjofcapra.net/publishers.html

Cassou, Michelle & Cubley, Steward

Life, Paint and Passion
Reclaiming the Magic of Spontaneous Expression
New York: P. Tarcher/Putnam, 1996

Castaneda, Carlos

The Teachings of Don Juan
A Yaqui Way of Knowledge
Washington: Square Press, 1985

Journey to Ixtlan
Washington: Square Press: 1991

Tales of Power
Washington: Square Press, 1991

The Second Ring of Power
Washington: Square Press, 1991

Castel, Robert

L'ordre psychiatrique, l'âge d'or de l'aliénisme
Paris: Éditions de Minuit, 1977

Cayce, Edgar
Modern Prophet
Four Complete Books
'Edgar Cayce On Prophecy'
'Edgar Cayce On Religion and Psychic Experience'
'Edgar Cayce On Mysteries of the Mind'
'Edgar Cayce On Reincarnation'
By Mary Ellen Carter
Ed. by Hugh Lynn Cayce
New York: Random House, 1968

Chaplin, Charles
My Autobiography
New York: Plume, 1992
Originally published in 196

Chevalier, Jean & Gheerbrant, Alain
A Dictionary of Symbols
Translated from the French by John Buchanan-Brown
New York: Penguin, 1996

Cho, Susanne
Kindheit und Sexualität im Wandel der Kulturgeschichte
Eine Studie zur Bedeutung der kindlichen Sexualität unter besonderer Berücksichtigung des 17. und 20. Jahrhunderts
Zürich, 1983 (Doctoral thesis)

Chopra, Deepak
Creating Affluence
The A-to-Z Steps to a Richer Life
New York: Amber-Allen Publishing (2003)

Life After Death
The Book of Answers
London: Rider, 2006

Leben nach dem Tod
Das letzte Geheimnis unserer Existenz
Berlin: Allegria Verlag, 2008

Synchrodestiny
Discover the Power of Meaningful Coincidence to Manifest Abundance
Audio Book / CD
Niles, IL: Nightingale-Conant, 2006

The Seven Spiritual Laws of Success
A Practical Guide to the Fulfillment of Your Dreams
Audio Book / CD
New York: Amber-Allen Publishing (2002)

Die Sieben Geistigen Gesetze des Erfolgs
Berlin: Ullstein Verlag, 2004

The Spontaneous Fulfillment of Desire
Harnessing the Infinite Power of Coincidence
New York: Random House Audio, 2003

Cicero, Marcus Tullius

Selected Works
New York: Penguin, 1960 (Penguin Classics)

Clarke, Ronald

Einstein: The Life and Times
New York: Avon Books, 1970

Clarke-Steward, S., Friedman, S. & Koch, J.

Child Development, A Topical Approach
London: John Wiley, 1986

Cleary, Thomas

The Taoist I Ching
Translated by Thomas Cleary
Boston & London: Shambhala, 1986

Constantine, Larry L.

Children & Sex
New Findings, New Perspectives
Larry L. Constantine & Floyd M. Martinson (Eds.)
Boston: Little, Brown & Company, 1981

Treasures of the Island
Children in Alternative Lifestyles
Beverly Hills: Sage Publications, 1976

Where are the Kids?
in: Libby & Whitehurst (ed.)
Marriage and Alternatives
Glenview: Scott Foresman, 1977

Open Family
A Lifestyle for Kids and other People
26 FAMILY COORDINATOR 113-130 (1977)

Cook, M. & Howells, K. (Eds.)

Adult Sexual Interest in Children
Academic Press, London, 1980

Coudenhove-Kalergi, Richard N.

Paneuropa
Wien-Leipzig: Paneuropa Verlag, 1926

Covey, Stephen R.

The 7 Habits of Highly Effective People
Powerful Lessons in Personal Change
New York: Free Press, 2004
15th Anniversary Edition
First Published in 1989

Die 7 Wege zur Effektivität
Prinzipien für persönlichen und beruflichen Erfolg
Offenbach: Gabal Verlag, 2009

The 8th Habit
From Effectiveness to Greatness
London: Simon & Schuster, 2004

Der 8. Weg
Von der Effektivität zur wahren Grösse
Offenbach: Gabal Verlag, 2006

Covitz, Joel
Emotional Child Abuse
The Family Curse
Boston: Sigo Press, 1986

Cox, Geraldine
The Home is Where the Heart is
Sydney: Macmillan, 2000

Craze, Richard
Feng Shui
Feng Shui Book & Card Pack
London: Thorsons, 1997

Cross, Sir Rupert
Cross on Evidence
5th ed.
London: Butterworths, 1979

Introduction to Criminal Law
10th Edition
London: Butterworths, 1984

Currier, Richard L.
Juvenile Sexuality in Global Perspective
in : Children & Sex, New Findings, New Perspectives
Larry L. Constantine & Floyd M. Martinson (Eds.)
Boston: Little, Brown & Company, 1981

D

Daco, Pierre
Les triomphes de la psychanalyse de Pierre Daco
Bruxelles: Éditions Gérard & Co., 1965 (Marabout)

Dalai Lama
Ethics for the New Millennium
New York: Penguin Putnam, 1999

David-Neel, Alexandra
Magic and Mystery in Tibet
New York: Dover Publications, 1971

The Secret Oral Teachings in Tibetan Buddhist Sects
New York: Secrets of Light Publishers, 1981

Initiations and Initiates in Tibet
New York: Dover Publications, 1993

Immortality and Reincarnation
Wisdom from the Forbidden Journey
New York: Inner Tradition, 1997

Davidson, Gustav
A Dictionary of Angels
Including Fallen Angels
New York: Free Press, 1967

Davis, A. J.
Sexual Assaults in the Philadelphia Prison System and Sheriff's Van
Trans-Action 6, 2, 8-16 (1968)

Dean & Bruyn-Kops
The Crime and the Consequences of Rape
New York: Thomas, 1982

De Bono, Edward

The Use of Lateral Thinking
New York: Penguin, 1967

The Mechanism of Mind
New York: Penguin, 1969

Sur/Petition
London: HarperCollins, 1993

Tactics
London: HarperCollins, 1993
First published in 1985

Taktiken und Strategien erfolgreicher Menschen
Frankfurt/M: MVG Verlag, 1995

Serious Creativity
Using the Power of Lateral Thinking to Create New Ideas
London: HarperCollins, 1996

Delacour, Jean-Baptiste

Glimpses of the Beyond
New York: Bantam Dell, 1975

Deleuze, Gilles, Guattari, Felix

L'Anti-Oedipe
Capitalisme et Schizophrénie
Nouvelle Édition Augmentée
Paris: Éditions de Minuit, 1973

DeMause, Lloyd

The History of Childhood
New York, 1974

Foundations of Psychohistory
New York: Creative Roots, 1982

DeMeo, James

Heretic's Notebook
Emotions, Protocells, Ether-Drift and Cosmic Life Energy
with New Research Supporting Wilhelm Reich
Pulse of the Planet, #5 (2002)
Ashland, Oregon: Orgone Biophysical Research Laboratories, Inc., 2002

Nach Reich, Neue Forschungen zur Orgonomie
Sexualökonomie / Die Entdeckung der Orgonenergie
Herausgegeben zusammen mit Professor Bernd Senf, Berlin
Frankfurt/M: Zweitausendeins Verlag, 1997

Saharasia
The 4000 BCE Origins of Child Abuse, Sex-Repression,
Warfare and Social Violence in the Deserts of the Old World
Ashland, Oregon: Orgone Biophysical Research Laboratories, Inc., 1998

Deshimaru, Taisen

Zen et vie quotidienne
Paris: Albin Michel, 1985

Diamond, Stephen A., May, Rollo

Anger, Madness, and the Daimonic
The Psychological Genesis of Violence, Evil and Creativity
New York: State University of New York Press, 1999

DiCarlo, Russell E. (Ed.)

Towards A New World View
Conversations at the Leading Edge
Erie, PA: Epic Publishing, 1996

Dicta et Françoise

Tarot de Marseille
Paris: Mercure de France, 1980

Dolto, Françoise

La Cause des Enfants
Paris: Laffont, 1985

Mein Leben auf der Seite der Kinder
Ein Plädoyer für eine kindgerechte Welt
Hamburg: Lübbe Verlagsgruppe, 1993

Psychanalyse et Pédiatrie
Paris: Seuil, 1971

Psychoanalyse und Kinderheilkunde
Frankfurt/M: Suhrkamp, 1997

Séminaire de Psychanalyse d'Enfants, 1
Paris: Seuil, 1982

Séminaire de Psychanalyse d'Enfants, 2
Paris: Seuil, 1985

Séminaire de Psychanalyse d'Enfants, 3
Paris: Seuil, 1988

Praxis der Kinderanalyse. Ein Seminar.
Hamburg: Klett-Cotta, 1985

Alles ist Sprache
Kindern mit Worten helfen
Berlin: Quadriga, 1996

Über das Begehren
Die Anfänge der menschlichen Kommunikation
2. Auflage
Hamburg: Klett-Cotta, 1996

Kinder stark machen
Die ersten Lebensjahre
Berlin: Beltz Verlag, 2000

L'évangile au risque de la psychanalyse
Paris: Seuil, 1980

Dover, K.J.
Greek Homosexuality
New York: Fine Communications, 1997

Dreher & Tröndle
Strafgesetzbuch und Nebengesetze
42. Aufl.
München: Beck, 1985

Dürckheim, Karlfried Graf
Hara: The Vital Center of Man
Rochester: Inner Traditions, 2004

Hara
Die Erdmitte des Menschen
Neuausgabe
München: O.W. Barth bei Scherz, 2005

Zen and Us
New York: Penguin Arkana 1991

The Call for the Master
New York: Penguin Books, 1993

Absolute Living
The Otherworldly in the World and the Path to Maturity
New York: Penguin Arkana, 1992

The Way of Transformation
Daily Life as a Spiritual Exercise
London: Allen & Unwin, 1988

Der Alltag als Übung
Vom Weg der Verwandlung
Bern: Huber, 2008

The Japanese Cult of Tranquility
London: Rider, 1960

Kultur der Stille
Frankfurt/M: Weltz Verlag, 1997

E

Eden, Donna & Feinstein, David

Energy Medicine
New York: Tarcher/Putnam, 1998

The Energy Medicine Kit
Simple Effective Techniques to Help You Boost Your Vitality
Boulder, Co.: Sounds True Editions, 2004

The Promise of Energy Psychology
With David Feinstein and Gary Craig
Revolutionary Tools for Dramatic Personal Change
New York: Jeremy P. Tarcher/Penguin, 2005

Edmunds, Francis

An Introduction to Anthroposophy
Rudolf Steiner's Worldview
London: Rudolf Steiner Press, 2005

Edwardes, A.

The Jewel of the Lotus
New York, 1959

Einstein, Albert

The World As I See It
New York: Citadel Press, 1993

Mein Weltbild
Berlin: Ullstein, 2005

Out of My Later Years
New York: Outlet, 1993

Ideas and Opinions
New York: Bonanza Books, 1988

Einstein sagt
Zitate, Einfälle, Gedanken
München: Piper, 2007

Albert Einstein Notebook
London: Dover Publications, 1989

Eisler, Riane

The Chalice and the Blade
Our history, Our future
San Francisco: Harper & Row, 1995

Kelch und Schwert, Unsere Geschichte, unsere Zukunft
Weibliches und männliches Prinzip in der Geschichte
Freiburg: Arbor Verlag, 2005

Sacred Pleasure: Sex, Myth and the Politics of the Body
New Paths to Power and Love
San Francisco: Harper & Row, 1996

The Partnership Way
New Tools for Living and Learning
With David Loye
Brandon, VT: Holistic Education Press, 1998

The Real Wealth of Nations
Creating a Caring Economics
San Francisco: Berrett-Koehler Publishers, 2008

Eliade, Mircea

Shamanism
Archaic Techniques of Ecstasy
New York: Pantheon Books, 1964

Ellis, Havelock

Sexual Inversion
Republished
New York: University Press of the Pacific, 2001
Originally published in 1897

Analysis of the Sexual Impulse
Love and Pain
The Sexual Impulse in Women
Republished
New York: University Press of the Pacific, 2001
Originally published in 1903

The Dance of Life
New York: Greenwood Press Reprint Edition, 1973
Originally published in 1923

Elwin, V.

The Muria and their Ghotul
Bombay: Oxford University Press, 1947

Emerson, Ralph Waldo

The Essays of Ralph Waldo Emerson
Cambridge, Mass.: Harvard University Press, 1987

Emoto, Masaru

The Hidden Messages in Water
New York: Atria Books, 2004

Die Botschaft des Wassers
Burgrain: Koha Verlag, 2008

The Secret Life of Water
New York: Atria Books, 2005

Die Heilkraft des Wassers
Burgrain: Koha Verlag, 2004

Encyclopédies d'Aujourd'hui
Encyclopédie de la Franc-Maçonnerie
Paris: Librairie Générale Française, 2000
(La Pochothèque)

Erickson, Milton H.
My Voice Will Go With You
The Teaching Tales of Milton H. Erickson
by Sidney Rosen (Ed.)
New York: Norton & Co., 1991

Complete Works 1.0, CD-ROM
New York: Milton H. Erickson Foundation, 2001

Erikson, Erik H.
Childhood and Society
New York: Norton, 1993
First published in 1950

Erman/Ranke
Ägypten und Ägyptisches Leben im Altertum
Hildesheim: Gerstenberg, 1981

Evans-Wentz, Walter Yeeling
The Fairy Faith in Celtic Countries
London: Frowde, 1911
Republished by Dover Publications
(Minneola, New York), 2002

F

Farson, Richard
Birthrights
A Bill of Rights for Children
Macmillan, New York, 1974

Feinberg, Joel
Harmless Wrongdoing
The Moral Limits of the Criminal Law, Vol. 4
New York: Oxford University Press, 1990

Fensterhalm, Herbert
Don't Say Yes When You Want to Say No
With Jean Bear
New York: Dell, 1980

Fericla, Josep M.
Al trasluz de la Ayahuasca
Antropología cognitiva, oniromancia y consciencias alternativas
Barcelona: La Liebre de Marzo, 2002

Finkelhor, David
Sexually Victimized Children
New York: Free Press, 1981

Finkelstein, Haim N. (Ed.)
The Collected Writings of Salvador Dali
Cambridge: Cambridge University Press, 1998

Flack, Audrey
Art & Soul
Notes on Creating
New York: E P Dutton, Reissue Edition, 1991

Forte, Robert (Ed.)
Entheogens and the Future of Religion
Council on Spiritual Practices, 2nd ed., 2000

Fortune, Mary M.
Sexual Violence
New York: Pilgrim Press, 1994

Foster/Freed

A Bill of Rights for Children
6 FAMILY LAW QUARTERLY 343 (1972)

Foucault, Michel

The History of Sexuality, Vol. I : The Will to Knowledge
London: Penguin, 1998
First published in 1976

The History of Sexuality, Vol. II : The Use of Pleasure
London: Penguin, 1998
First published in 1984

The History of Sexuality, Vol. III : The Care of Self
London: Penguin, 1998
First published in 1984

Fourcade, Jean-Michel

Analyse transactionnelle et bioénergie
Paris: Delarge, 1981

Foxwood, Orion

The Faery Teachings
Arcata, CA: R.J. Steward Books, 2007

Franz Anton Mesmer

Franz Anton Mesmer und die Geschichte des Mesmerismus
Beiträge zum internationalen wissenschaftlichen Symposium anlässlich des 250. Geburtstages von Mesmer
Stuttgart, 1985

Freud, Anna

War and Children
London: 1943

Freud, Sigmund

Three Essays on the Theory of Sexuality
in: The Standard Edition of the Complete Psychological
Works of Sigmund Freud
London: Hogarth Press, 1953-54
Vol. 7, pp. 130 ff
(first published in 1905)

Drei Abhandlungen zur Sexualtheorie
Frankfurt/M: Fischer, 1991

The Interpretation of Dreams
New York: Avon, Reissue Edition, 1980
and in: The Standard Edition of the Complete Psychological
Works of Sigmund Freud , (24 Volumes) ed. by James Strachey
New York: W. W. Norton & Company, 1976

Die Traumdeutung
Frankfurt/M: Fischer, 2005

Totem and Taboo
New York: Routledge, 1999
Originally published in 1913

Totem und Tabu
Einige Übereinstimmungen im Seelenleben der Wilden und
der Neurotiker
Frankfurt/M: Fischer Verlag, 1972

Freund, Kurt

Assessment of Pedophilia
in: Cook, M. and Howells, K. (eds.)
Adult Sexual Interest in Children
Academic Press, London, 1980

Frisch, Max

Biedermann und die Brandstifter
München: Suhrkamp, 1996
Erstmals 1955 als Hörspiel veröffentlicht

Fromm, Erich

The Anatomy of Human Destructiveness
New York: Owl Book, 1992
Originally published in 1973

Anatomie der menschlichen Destruktivität
Berlin: Rowohlt, 1977

Escape from Freedom
New York: Owl Books, 1994
Originally published in 1941

Die Furcht vor der Freiheit
München: DTV Verlag, 1993

To Have or To Be
New York: Continuum International Publishing, 1996
Originally published in 1976

Haben oder Sein
Die seelischen Grundlagen einer neuen Gesellschaft
München: DTV Verlag, 2005

The Art of Loving
New York: HarperPerennial, 2000
Originally published in 1956

Die Kunst des Liebens
Berlin: Ullstein, 2005

G

Gates, Bill

The Road Ahead
New York, Penguin, 1996
(Revised Edition)

Gawain, Shakti
Creative Visualization
Use the Power of Your Imagination to Create What You Want
Novato, CA: New World Library, 1995

Creative Visualization Meditations (Reader)
Novato, CA: New World Library, 1997

Geldard, Richard
Remembering Heraclitus
New York: Lindisfarne Books, 2000

Gerber, Richard
A Practical Guide to Vibrational Medicine
Energy Healing and Spiritual Transformation
New York: Harper & Collins, 2001

Geller, Uri
The Mindpower Kit
Includes Book, Audiotape, Quartz Crystal And Meditation Circle
New York: Penguin, 1996

Gesell, Izzy
Playing Along
37 Group Learning Activities Borrowed from Improvisational Theater
Whole Person Associates, 1997

Ghiselin, Brewster (Ed.)
The Creative Process
Reflections on Invention in the Arts and Sciences
Berkeley: University of California Press, 1985
First published in 1952

Gibson, Ian
The Shameful Life of Salvador Dali
New York: Norton, 1998

Gil, David G.
Societal Violence and Violence in Families
in: David G. Gil, Child Abuse and Violence
New York: Ams Press, 1928

Gimbutas, Marija
The Language of the Goddess
London: Thames & Hudson, 2001

Glucksmann, André, Wolton, Thierry
Silence On Tue
Paris: Grasset, 1986

Goethe, Johann Wolfgang von
The Theory of Colors
New York: MIT Press, 1970
First published in 1810

Goethes Farbenlehre
Leipzig: Seemann-Henschel Verlag, 1998

Goldenstein, Joyce
Einstein: Physicist and Genius
(Great Minds of Science)
New York: Enslow Publishers, 1995

Goldman, Jonathan & Goldman, Andi
Tantra of Sound
Frequencies of Healing
Charlottesville: Hampton Roads, 2005

Tantra des Klanges
Mehr Liebe und Intimität in der Partnerschaft
Mit CD
Hanau: Amra Verlag, 2009

Healing Sounds
The Power of Harmonies
Rochester: Healing Arts Press, 2002

Klangheilung
Die Schöpferkraft des Obertongesangs
Hanau: Amra Verlag, 2008

Healing Sounds
Principles of Sound Healing
DVD, 90 min.
Sacred Mysteries, 2004

Goldstein, Jeffrey H.
Aggression and Crimes of Violence
New York, 1975

Goleman, Daniel
Emotional Intelligence
New York, Bantam Books, 1995

EQ. Emotionale Intelligenz
München: DTV Verlag, 1997

Goodwin, Matthew O.
The Complete Numerology Guide
New York: Red Wheel/Weiser, 1988

Gordon, Rosemary
Pedophilia: Normal and Abnormal
in: Kraemer, The Forbidden Love
London, 1976

Goswami, Amit
The Self-Aware Universe
How Consciousness Creates the Material World
New York: Tarcher/Putnam, 1995

Das Bewusste Universum
Wie Bewusstsein die materielle Welt erschafft
Stuttgart: Lüchow Verlag, 2007

Gottlieb, Adam
Peyote and Other Psychoactive Cacti
Ronin Publishing, 2nd edition, 1997

Grant
Grant's Method of Anatomy
10th ed., by John V. Basmajian
Baltimore, London: Williams & Wilkins, 1980

Greene, Liz
Astrology of Fate
York Beach, ME: Red Wheel/Weiser, 1986

Saturn
A New Look at an Old Devil
York Beach, ME: Red Wheel/Weiser, 1976

The Astrological Neptune and the Quest for Redemption
Boston: Red Wheel Weiser, 1996

The Mythic Journey
With Juliet Sharman-Burke
The Meaning of Myth as a Guide for Life
New York: Simon & Schuster (Fireside), 2000

Die Mythische Reise
Die Bedeutung der Mythen als ein Führer durch das Leben
München: Atmosphären Verlag, 2004

The Mythic Tarot
With Juliet Sharman-Burke
New York: Simon & Schuster (Fireside), 2001
Originally published in 1986

Le Tarot Mythique
Une nouvelle approche du Tarot
Paris: Solar, 1988

The Luminaries
The Psychology of the Sun and Moon in the Horoscope
With Howard Sasportas
York Beach, ME: Red Wheel/Weiser, 1992

Sonne und Mond
Die Bedeutung der grossen Lichter in der Mythologie und im Horoskop
Saarbrücken: Neue Erde/Lentz, 2000

Greer, John Michael
Earth Divination, Earth Magic
A Practical Guide to Geomancy
New York: Llewellyn Publications, 1999

Grinspoon, Lester
Marihuana
The Forbidden Medicine
With James B. Bakalar
New Haven, CT: Yale University Press, 1997
First published in 1971

Groeben/Boeckh/Thiesing/Ehlermann
Kommentar zum EWG-Vertrag
Band 2, Dritte Auflage
Baden-Baden: Nomos, 1983

Grof, Stanislav
Ancient Wisdom and Modern Science
New York: State University of New York Press, 1984

Beyond the Brain
Birth, Death and Transcendence in Psychotherapy
New York: State University of New York, 1985

LSD: Doorway to the Numinous
The Groundbreaking Psychedelic Research into Realms of the
Human Unconscious
Rochester: Park Street Press, 2009

Psychologie transpersonnelle
Paris: Rocher, 1984

Realms of the Human Unconscious
Observations from LSD Research
New York: E.P. Dutton, 1976

The Cosmic Game
Explorations of the Frontiers of Human Consciousness
New York: State University of New York Press, 1998

The Holotropic Mind
The Three Levels of Human Consciousness
With Hal Zina Bennett
New York: HarperCollins, 1993

When the Impossible Happens
Adventures in Non-Ordinary Reality
Louisville, CO: Sounds True, 2005

Wir wissen mehr als unser Gehirn
Die Grenzen des Bewusstseins überschreiten
Freiburg: Herder, 2007

Groth, A. Nicholas

Men Who Rape
The Psychology of the Offender
New York: Perseus Publishing, 1980

Grout, Pam

Art & Soul
New York: Andrews McMeel Publishing, 2000

Gunn, John
Violence
New York/Washington, 1973

Gurdjieff, George Ivanovich
The Herald of Coming Good
London: Samuel Weiser, 1933

H

Hall, Manly P.
The Pineal Gland
The Eye of God
Article extracted from the book: Man the Grand Symbol of the Mysteries
Kessinger Publishing Reprint

The Secret Teachings of All Ages
Reader's Edition
New York: Tarcher/Penguin, 2003
Originally published in 1928

Hameroff, Newberg, Woolf, Bierman et al.
Consciousness
20 Scientists Interviewed
Director: Gregory Alsbury
5 DVD Box Set, 540 min.
New York: Alsbury Films, 2003

Hargous, Sabine
Les appeleurs d'âmes
L'univers chamanique des Indiens des Andes
Paris: Albin Michel, 1985

Harner, Michael
Ways of the Shaman
New York: Bantam, 1982 (Originally published in 1980)

Der Weg des Schamanen
Das praktische Grundlagenbuch zum Schamanismus
Genf: Ariston, 2007

Chamane
Les secrets d'un sorcier indien d'Amérique du Nord
Paris: Albin Michel, 1982

Hasegawa, Tsuyoshi
Racing the Enemy
Stalin, Truman, and the Surrender of Japan
Cambridge, MA: Belknap Press of Harvard University Press, 2006

Henkin/Pugh/Schachter/Smit
International Law
Cases and Materials
St. Paul (West): American Casebook Series, 1980

Herman, Dean M.
A Statutory Proposal to Prohibit the Infliction of Violence upon Children
19 FAMILY LAW QUARTERLY, 1986, 1-52

Hermes Trismegistos
Corpus Hermeticum
New York: Edaf, 2001

Héroard, J.
Journal de Jean Héroard sur l'Enfance et la Jeunesse de Louis XIII
Paris: Soulié/Barthélemy, 1868

Herrigel, Eugen
Zen in the Art of Archery
New York: Vintage Books, 1999
Originally published in 1971

Hicks, Esther and Jerry
The Amazing Power of Deliberate Intent
Living the Art of Allowing
Carlsbad, CA: Hay House, 2006

Hobbes, Thomas
Leviathan (1651)
New York: Longman Library, 2006

Hofmann, Albert
LSD, My Problem Child
Reflections on Sacred Drugs, Mysticism and Science
Santa Cruz, CA: Multidisciplinary Association for Psychedelic Studies, 2009
Originally published in 1980

LSD, Mein Sorgenkind
Die Entdeckung der 'Wunderdroge'
München: DTV Verlag, 1999

Holmes, Ernst
The Science of Mind
A Philosophy, A Faith, A Way of Life
New York: Jeremy P. Tarcher/Putnam, 1998
First Published in 1938

Holstiege, Hildegard
Montessori Pädagogik und soziale Humanität
Freiburg: Herder, 1994

Hood, J. X.
Scientific Curiosities of Love, Sex and Marriage
A Survey of Sex Relations, Beliefs and Customs of Mankind in Different Countries and Ages
New York, 1951

Houston, Jean
The Possible Human
A Course in Enhancing Your Physical, Mental, and Creative Abilities
New York: Jeremy P. Tarcher/Putnam, 1982

Howells, Kevin
Adult Sexual Interest in Children
Considerations Relevant to Theories of Aetiology in:
Cook, M. and Howells, K. (eds.): Adult Sexual Interest in Children
Academic Press, London, 1980

Huang, Alfred
The Complete I Ching
The Definite Translation from Taoist Master Alfred Huang
Rochester, NY: Inner Traditions, 1998

Hunt, Valerie
Infinite Mind
Science of the Human Vibrations of Consciousness
Malibu, CA: Malibu Publishing, 2000

Huxley, Aldous
The Doors of Perception and Heaven and Hell
London: HarperCollins (Flamingo), 1994
(originally published in 1954)

The Perennial Philosophy
San Francisco: Harper & Row, 1970

I

Innocenti Declaration
Declaration on the Protection, Promotion and Support of Breastfeeding
http://www.innocenti15.net/inno.htm

J

Jackson, Nigel
The Rune Mysteries
With Silver RavenWolf
St. Paul, Minn.: Llewellyn Publications, 2000

Jackson, Stevi
Childhood and Sexuality
New York: Blackwell, 1982

Jaffe, Hans L.C.
Picasso
New York: Abradale Press, 1996

James, William
Writings 1902-1910
The Varieties of Religious Experience / Pragmatism / A Pluralistic Universe / The Meaning of Truth / Some Problems of Philosophy / Essays
New York: Library of America, 1988

Jampolsky, Gerald
Aimer c'est se libérer de la peur
Genève: Éditions Soleil, 1986

Janov, Arthur
Primal Man
The New Consciousness
New York: Crowell, 1975

Das Neue Bewusstsein
Frankfurt/M: Fischer Verlag, 1988
Urausgabe 1975

Johnson, Paul
A History of the Jews
New York: Harper & Row, 1987

Johnston & Deisher
Contemporary Communal Child Rearing: A First Analysis
52 PEDIATRICS 319 (1973)

Jones, W.H.S., Litt, D.
Pliny Natural History
Cambridge, Mass.: Harvard University Press, 1980

Jung, Carl Gustav
Archetypen
München: DTV Verlag, 2001

Archetypes of the Collective Unconscious
in: The Basic Writings of C.G. Jung
New York: The Modern Library, 1959, 358-407

Collected Works
New York, 1959

Dialectique du moi et de l'inconscient
Paris, Gallimard, 1991

On the Nature of the Psyche
in: The Basic Writings of C.G. Jung
New York: The Modern Library, 1959, 47-133

Psychological Types
Collected Writings, Vol. 6
Princeton: Princeton University Press, 1971

Psychologie und Religion
München: DTV Verlag, 2001

Psychology and Religion
in: The Basic Writings of C.G. Jung
New York: The Modern Library, 1959, 582-655

Religious and Psychological Problems of Alchemy
in: The Basic Writings of C.G. Jung
New York: The Modern Library, 1959, 537-581

Symbol und Libido
Freiburg: Walter Verlag, 1987

Synchronizität, Akausalität und Okkultismus
Frankfurt/M: DTV, 2001

The Basic Writings of C.G. Jung
New York: The Modern Library, 1959

The Development of Personality
Collected Writings, Vol. 17
Princeton: Princeton University Press, 1954

The Meaning and Significance of Dreams
Boston: Sigo Press, 1991

The Myth of the Divine Child
in: Essays on A Science of Mythology
Princeton, N.J.: Princeton University Press Bollingen Series XXII, 1969. (With Karl Kerenyi)

Traum und Traumdeutung
München: DTV Verlag, 2001

Two Essays on Analytical Psychology
Collected Writings, Vol. 7
Princeton: Princeton University Press, 1972
First published by Routledge & Kegan Paul, Ltd., 1953

Zur Psychologie westlicher und östlicher Religion
Fünfte Auflage
Olten: Walter Verlag, 1988

K

Kahn, Charles (Ed.)
The Art and Thought of Heraclitus
Cambridge: Cambridge University Press, 2008

Kaiser, Edmond
La Marche aux Enfants
Lausanne: P.-M. Favre, 1979

Kalweit, Holger
Shamans, Healers and Medicine Men
Boston and London: Shambhala, 1992
Originally published with Kösel Verlag, Munich, in 1987

Kant, Immanuel
Kant's Werke
Band VIII, Abhandlungen nach 1781 (Neudruck)
Berlin und Leipzig: Walter de Gruyter, 1923

Kapleau, Roshi Philip
Three Pillars of Zen
Boston: Beacon Press, 1967

Karagulla, Shafica
The Chakras
Correlations between Medical Science and Clairvoyant Observation
With Dora van Gelder Kunz
Wheaton: Quest Books, 1989

Die Chakras und die feinstofflichen Körper des Menschen
Mit Dora van Gelder-Kunz
Grafing: Aquamarin Verlag, 1994

Karremann, Manfred

Es geschieht am helllichten Tag
Die Verborgene Welt der Pädophilen
und wie wir unsere Kinder vor Missbrauch Schützen
Köln: Dumont, 2007

Kerner Justinus

F.A. Mesmer aus Schwaben
Frankfurt/M, 1856

Kiang, Kok Kok

The I Ching
An Illustrated Guide to the Chinese Art of Divination
Singapore: Asiapac, 1993

Kiesewetter, Carl

Franz Anton Mesmer's Leben und Lehre
Leipzig, 1893

Kingston, Karen

Creating Sacred Space With Feng Shui
New York: Broadway Books, 1997

Kinski, Klaus

Kinski Uncut: The Autobiography of Klaus Kinski
New York: Penguin, 1997

Klein, Melanie

Love, Guilt and Reparation, and Other Works 1921-1945
New York: Free Press, 1984
(Reissue Edition)

Envy and Gratitude and Other Works 1946-1963
New York: Free Press, 2002
(Reissue Edition)

Klimo, Jon
Channeling
Investigations on Receiving Information from Paranormal Sources
New York: North Atlantic Books, 1988

Koestler, Arthur
The Act of Creation
New York: Penguin Arkana, 1989.
Originally published in 1964

Kraemer
The Forbidden Love
London, 1976

Krafft-Ebing, Richard von
Psychopathia sexualis
New York: Bell Publishing, 1965
Originally published in 1886

Krause, Donald G.
The Art of War for Executives
London: Nicholas Brealey Publishing, 1995

Krishnamurti, J.
Freedom From The Known
San Francisco: Harper & Row, 1969

The First and Last Freedom
San Francisco: Harper & Row, 1975

Education and the Significance of Life
London: Victor Gollancz, 1978

Commentaries on Living
First Series
London: Victor Gollancz, 1985

Commentaries on Living
Second Series
London: Victor Gollancz, 1986
Krishnamurti's Journal
London: Victor Gollancz, 1987

Krishnamurti's Notebook
London: Victor Gollancz, 1986

Beyond Violence
London: Victor Gollancz, 1985

Beginnings of Learning
New York: Penguin, 1986

The Penguin Krishnamurti Reader
New York: Penguin, 1987

On God
San Francisco: Harper & Row, 1992

On Fear
San Francisco: Harper & Row, 1995

The Essential Krishnamurti
San Francisco: Harper & Row, 1996

The Ending of Time
With Dr. David Bohm
San Francisco: Harper & Row, 1985

Kwok, Man-Ho

The Feng Shui Kit
London: Piatkus, 1995

L

Labate, Beatriz Caluby

Ayahuasca Religions
A Comprehensive Bibliography and Critical Essays
Santa Cruz, CA: Maps, 2009

Laing, Ronald David

Divided Self
New York: Viking Press, 1991

R.D. Laing and the Paths of Anti-Psychiatry
ed., by Z. Kotowicz
London: Routledge, 1997

The Politics of Experience
New York: Pantheon, 1983

Sagesse, déraison et folie
Paris: Seuil, 1986

Lakhovsky, Georges

La Science et le Bonheur
Longévité et Immortalité par les Vibrations
Paris: Gauthier-Villars, 1930

Le Secret de la Vie
Paris: Gauthier-Villars, 1929

Secret of Life
New York: Kessinger Publishing, 2003

L'étiologie du Cancer
Paris: Gauthier-Villars, 1929

L'Universion
Paris: Gauthier-Villars, 1927

Lanouette, William

Genius in the Shadows
A Biography of Leo Szilard, the Man behind the Bomb
With Bela Silard
Chicago: University of Chicago Press, 1994

Laszlo, Ervin

Holos. Die Welt der neuen Wissenschaften
Petersberg: Via Nova Verlag, 2002

Science and the Akashic Field
An Integral Theory of Everything
Rochester: Inner Traditions, 2004

Macroshift
Die Herausforderung
Frankfurt/M: Insel Verlag, 2003

Quantum Shift to the Global Brain
How the New Scientific Reality Can Change Us and Our World
Rochester: Inner Traditions, 2008

Science and the Reenchantment of the Cosmos
The Rise of the Integral Vision of Reality
Rochester: Inner Traditions, 2006

The Akashic Experience
Science and the Cosmic Memory Field
Rochester: Inner Traditions, 2009

The Chaos Point
The World at the Crossroads
Newburyport, MA: Hampton Roads Publishing, 2006

Laud, Anne & Gilstrop, May
Violence in the Family
A Selected Bibliography on Child Abuse, Sexual Abuse of Children & Domestic Violence
June 1985
University of Georgia Libraries
Bibliographical Series, No. 32

Lauterpacht, E., Q.C.
International Law Reports
Cambridge: Grotius Publishers

Lauterpacht, Hersch
International Law
Ed. By E. Lauterpacht, Q.C.
Vol. 3
London: Cambridge University Press, 1977

LaViolette, Paul A.
Secrets of Antigravity Propulsion: Tesla, UFOs, and Classified Aerospace Technology
New York: Bear & Company, 2008

The U.S. Antigravity Squadron
In: Thomas Valone, Ed., *Electrogravitics Systems, Reports on a New Propulsion Methodology*
Washington, D.C.: Integrity Research Institute, 1993, 78-96

Leadbeater, Charles Webster
Astral Plane
Its Scenery, Inhabitants and Phenomena
Kessinger Publishing Reprint Edition, 1997

Dreams
What they Are and How they are Caused
London: Theosophical Publishing Society, 1903
Kessinger Publishing Reprint Edition, 1998

The Inner Life
Chicago: The Rajput Press, 1911
Kessinger Publishing

Leary, Timothy

Our Brain is God
Berkeley, CA: Ronin Publishing, 2001
Author Copyright 1988

Über die Kriminalisierung des Natürlichen
Löhrbach: Werner Pieper Verlag, 1990

Leboyer, Frederick

Birth Without Violence
New York, 1975

Pour une Naissance sans Violence
Paris: Seuil, 1974

Geburt ohne Gewalt
München: Kösel 1981

Cette Lumière d'où vient l'Enfant
Paris: Seuil, 1978

Inner Beauty, Inner Light
New York: Newmarket Press, 1997

Weg des Lichts
München: Kösel, 1991

Loving Hands
The Traditional Art of Baby Massage
New York: Newmarket Press, 1977

Sanfte Hände
Die Kunst der indischen Baby-Massage
München: Kösel, 1979

The Art of Breathing
New York: Newmarket Press, 1991

Le Crapouillot
Les pédophiles
Nouvelle série, n°73, Janvier 1984
Vincent Acker, Le Vilain Manège du Coral, pp. 36-42

LeCron, Leslie M.
L'auto-hypnose
8e édition
Genève: Ariston, 1984

Leggett, Trevor P.
A First Zen Reader
Rutland: C.E. Tuttle, 1980
Originally published in 1972

Lenihan, Eddie
Meeting the Other Crowd
The Fairy Stories of Hidden Ireland
With Carolyn Eve Green
New York: Jeremy P. Tarcher/Penguin, 2004
Authors Copyright 2003

Leonard, George, Murphy, Michael
The Live We Are Given
A Long Term Program for Realizing the
Potential of Body, Mind, Heart and Soul
New York: Jeremy P. Tarcher/Putnam, 1984

Leopardi, Angelo (Hrsg.)
Der Pädosexuelle Komplex
Frankfurt/M: Foerster Verlag, 1988

Licht, Hans
Sexual Life in Ancient Greece
New York: AMS Press, 1995

Liedloff, Jean
Continuum Concept
In Search of Happiness Lost
New York: Perseus Books, 1986
First published in 1977

Auf der Suche nach dem verlorenen Glück
Gegen die Zerstörung der Glücksfähigkeit in der frühen Kindheit
München: C.H. Beck Verlag, 2006

Lip, Evelyn
The Design & Feng Shui of Logos, Trademarks and Signboards
Singapore: Prentice Hall, 1995

Lipgens, Walter
Europa-Föderationspläne der Widerstandsbewegungen 1940-1945
München, 1968

Lipton, Bruce
The Biology of Belief
Unleashing the Power of Consciousness, Matter and Miracles
Santa Rosa, CA: Mountain of Love/Elite Books, 2005

Intelligente Zellen
Wie Erfahrungen unsere Gene steuern
Burgrain: Koha Verlag, 2006

Liss, Jérôme
Débloquez vos émotions
Lausanne: Éditions Far, 1988

Locke, John
Some Thoughts Concerning Education
London, 1690
Reprinted in: The Works of John Locke, 1823
Vol. IX., pp. 6-205

Gedanken über Erziehung
Ditzingen: Reclam Verlag, 1986

Long, Max *Freedom*
The Secret Science at Work
The Huna Method as a Way of Life
Marina del Rey: De Vorss Publications, 1995
Originally published in 1953

Geheimes Wissen hinter Wundern
Die Entdeckung der HUNA-Lehre
Darmstadt: Schirner Verlag, 2006

Growing Into Light
A Personal Guide to Practicing the Huna Method,
Marina del Rey: De Vorss Publications, 1955

Lowen, Alexander

Angst vor dem Leben
Über den Ursprung seelischen Leides und den Weg zu einem reicheren Dasein
München: Goldmann Wilhelm, 1989

Bioenergetics
New York: Coward, McGoegham 1975

Bioenergetik
Therapie der Seele durch Arbeit mit dem Körper
Berlin: Rowohlt, 2008

Depression and the Body
The Biological Basis of Faith and Reality
New York: Penguin, 1992

Fear of Life
New York: Bioenergetic Press, 2003

Honoring the Body
The Autobiography of Alexander Lowen
New York: Bioenergetic Press, 2004

Joy
The Surrender to the Body and to Life
New York: Penguin, 1995

Liebe und Orgasmus
Persönlichkeitserfahrung durch sexuelle Erfüllung
München: Goldmann Wilhelm, 1993

Love and Orgasm
New York: Macmillan, 1965

Love, Sex and Your Heart
New York: Bioenergetics Press, 2004

Narcissism: Denial of the True Self
New York: Macmillan, Collier Books, 1983

Narzissmus
Die Verleugnung des wahren Selbst
München: Goldmann Wilhelm, 1992

Pleasure: A Creative Approach to Life
New York: Bioenergetics Press, 2004
First published in 1970

The Language of the Body
Physical Dynamics of Character Structure
New York: Bioenergetics Press, 2006

Luna, Luis Eduardo & Amaringo, Pablo
Ayahuasca Visions
North Atlantic Books, 1999

Lusk, Julie T. (Editor)
30 Scripts for Relaxation Imagery & Inner Healing
Whole Person Associates, 1992

Lutyens, Mary
Krishnamurti: The Years of Fulfillment
New York: Avon Books, 1983

Krishnamurti: Die Biographie
München: Aquamarin Verlag, 1997

The Life and Death of Krishnamurti
Chennai: Krishnamurti Foundation India, 1990

Lutzbetak, Louis J.
Marriage and the Family in Caucasia
Vienna, 1951, first reprinting, 1966

M

Machiavelli, Niccolo
The Prince
New York: Soho Books, 2009
Written in 1513
First posthumous publishing 1531

Der Fürst
Frankfurt/M: Insel Verlag, 2009

Mack, Carol K. & Mack, Dinah
A Field Guide to Demons, Fairies, Fallen Angels, and Other Subversive Spirits
New York: Owl Books, 1998

Maharshi, Ramana
The Collected Works of Ramana Maharshi
New York: Sri Ramanasramam, 2002

The Essential Teachings of Ramana Maharshi
A Visual Journey
New York: Inner Directions Publishing, 2002
by Matthew Greenblad

Sei was du bist!
München: O.W. Barth, 2001

Nan Yar? Wer bin ich?
München: Kamphausen, 2002

Maisel, Eric

Fearless Creating
A Step-By-Step Guide to Starting and Completing
Work of Art
New York: Tarcher & Putnam, 1995

Malachi, Tau

Gnosis of the Cosmic Christ
A Gnostic Christian Kabbalah
St. Paul: Llewellyn Publications, 2005

Malinowski, Bronislaw

Crime und Custom in Savage Society
London: Kegan, 1926

Sex and Repression in Savage Society
London: Kegan, 1927

The Sexual Life of Savages in North West Melanesia
New York: Halycon House, 1929

Das Geschlechtsleben der Wilden in Nordwest-Melanesien
Liebe, Ehe und Familienleben bei den Eingeborenen
der Trobriand Inseln,
Britisch-Neuguinea
Eschborn: Klotz Verlag, 2005

Mallet, Carl-Heinz

Das Einhorn bin ich
Das Bild des Menschen im Märchen
Hamburg: Hoffmann & Campe Verlag, 1982

Untertan Kind
Nachforschungen über Erziehung
München: Max Hueber Verlag, 1987

Mann, Edward W.

Orgone, Reich & Eros
Wilhelm Reich's Theory of Life Energy
New York: Simon & Schuster (Touchstone), 1973

Mann, Sally

At Twelve
Portraits of Young Women
New York: Aperture, 1988

Immediate Family
New York: Phaidon Press, 1993

Marciniak, Barbara

Bringers of the Dawn
Teachings from the Pleiadians
New York: Bear & Co., 1992

Boten des Neuen Morgens
Lehren von den Pleiaden
Freiburg: Hermann Bauer Verlag, 1995

Martinson, Floyd M.

Sexual Knowledge
Values and Behavior Patterns
St. Peter: Minn.: Gustavus Adolphus College, 1966

Infant and Child Sexuality
St. Peter: Minn.: Gustavus Adolphus College, 1973

The Quality of Adolescent Experiences
St. Peter: Minn.: Gustavus Adolphus College, 1974

The Child and the Family
Calgary, Alberta: The University of Calgary, 1980

The Sex Education of Young Children
in: Lorna Brown (Ed.), *Sex Education in the Eighties*
New York, London: Plenum Press, 1981, pp. 51 ff.

The Sexual Life of Children
New York: Bergin & Garvey, 1994

Children and Sex, Part II: Childhood Sexuality
in: Bullough & Bullough, Human Sexuality (1994)
Pp. 111-116

Master Lam Kam Chuen

The Way of Energy
Mastering the Chinese Art of Internal
Strength with Chi Kung Exercise
New York: Simon & Schuster (Fireside), 1991

Master Liang, Shou-Yu & Wu, Wen-Ching

Tai Chi Chuan
24 & 48 Postures With Martial Applications
Roslindale, Mass.: YMAA Publication Center, 1996

Masters, R.E.L.

Forbidden Sexual Behavior and Morality
New York, 1962

McCarey, William A.

In Search of Healing
Whole-Body Healing Through the Mind-Body-Spirit Connection
New York: Berkley Publishing, 1996

McCormick

McCormick on Evidence
by Edward W. Cleary, 3d ed.
Lawyers Edition (Homebook Series)
St. Paul: West, 1984

McKenna, Terence

The Archaic Revival
San Francisco: Harper & Row, 1992

Food of The Gods
A Radical History of Plants, Drugs and Human Evolution
London: Rider, 1992

Die Speisen der Götter
Berlin: Synergia/Syntropia, 1996

The Invisible Landscape
Mind Hallucinogens and the I Ching
New York: HarperCollins, 1993
(With Dennis McKenna)

True Hallucinations
Being the Account of the Author's Extraordinary
Adventures in the Devil's Paradise
New York: Fine Communications, 1998

McLeod, Kembrew

Freedom of Expression
Resistance and Repression in the Age of Intellectual Property
Minneapolis, MN: University of Minnesota Press, 2007

McNiff, Shaun

Art as Medicine
Boston: Shambhala, 1992

Art as Therapy
Creating a Therapy of the Imagination
Boston/London: Shambhala, 1992

Trust the Process
An Artist's Guide to Letting Go
New York: Shambhala Publications, 1998

McTaggart, Lynne

The Field
The Quest for the Secret Force of the Universe
New York: Harper & Collins, 2002

Mead, Margaret

Sex and Temperament in Three Primitive Societies
New York, 1935

Meadows, Donella H.

Thinking in Systems
A Primer
White River, VT: Chelsea Green Publishing, 2008

Mehta, Rohit

J. Krishnamurti and the Nameless Experience
A Comprehensive Discussion of J. Krishnamurti's Approach to Life
Delhi: Motilal Banarsidass Publishers, 2002

Méric, de, Philippe

Le Yoga sans postures
Paris: Livre de Poche, 1967

Merle, Roger & Vitu, André
Traité de Croit Criminel
Droit Pénal Spécial
Vol. II, par André Vitu
Paris: Editions Cujas, 1982

Merleau-Ponty, Maurice
Phenomenology of Perception
London: Routledge, 1995
Originally published 1945

Phénoménologie de la perception
Paris: Gallimard, 1945

Metzner, Ralph (Ed.)
Ayahuasca, Human Consciousness and the Spirits of Nature
ed. by Ralph Metzner, Ph.D
New York: Thunder's Mouth Press, 1999

The Psychedelic Experience
A Manual Based on the Tibetan Book of the Dead
With Timothy Leary and Richard Alpert
New York: Citadel, 1995

Miller, Alice
Four Your Own Good
Hidden Cruelty in Child-Rearing and the Roots of Violence
New York: Farrar, Straus & Giroux, 1983

Am Anfang war Erziehung
München: Suhrkamp Verlag, 2008
Erstmals publiziert im Jahre 1986

Pictures of a Childhood
New York: Farrar, Straus & Giroux, 1986

The Drama of the Gifted Child
In Search for the True Self
translated by Ruth Ward
New York: Basic Books, 1996

Das Drama des Begabten Kindes
Und die Suche nach dem wahren Selbst
München: Suhrkamp Verlag, 1983

Der gemiedene Schlüssel
München: Suhrkamp, 2007

Das verbannte Wissen
Frankfurt/M: Suhrkamp, 1988

Thou Shalt Not Be Aware
Society's Betrayal of the Child
New York: Noonday, 1998

Du Sollst Nicht Merken
Variationen über das Paradies-Thema
Neuauflage
München: Suhrkamp, 2005

The Political Consequences of Child Abuse
in: The Journal of Psychohistory 26, 2 (Fall 1998)

Miller, Mary & Taube, Karl

An Illustrated Dictionary of the Gods and Symbols of Ancient Mexico and the Maya
London: Thames & Hudson, 1993

Moll, Albert

The Sexual Life of the Child
New York: Macmillan, 1912
First published in German as
Das Sexualleben des Kindes, 1909

Monroe, Robert
Ultimate Journey
New York: Broadway Books, 1994

Monsaingeon, Bruno
Svjatoslav Richter
Notebooks and Conversations
Princeton: Princeton University Press, 2002

Richter
Écrits, conversations
Paris: Éditions Van de Velde, 1998

Richter The Enigma / L'Insoumis / Der Unbeugsame
NVC Arts 1998 (DVD)

Montagu, Ashley
Touching
The Human Significance of the Skin
New York: Harper & Row, 1978

Körperkontakt
8. Auflage
Stuttgart: Klett/Cotta, 1995

Monter, E. William
Witchcraft in France and Switzerland
Ithaca & London: Cornell University Press, 1976

Montessori, Maria
The Absorbent Mind
Reprint Edition
New York: Buccaneer Books, 1995
First published in 1973

Das Kreative Kind
Der absorbierende Geist
Freiburg: Herder, 2007

Moody, Raymond
The Light Beyond
New York: Mass Market Paperback (Bantam), 1989

Moore, Thomas
Care of the Soul
A Guide for Cultivating Depth and Sacredness in Everyday Life
New York: Harper & Collins, 1994

Die Seele Lieben
Tiefe und Spiritualität im täglichen Leben
München: Droemer Knaur, 1995

Moser, Charles Allen
DSM-IV-TR and the Paraphilias: an argument for removal
With Peggy J. Kleinplatz
Journal of Psychology and Human Sexuality 17 (3/4), 91-109 (2005)

Murdock, G.
Social Structure
New York: Macmillan, 1960

Murphy, Joseph
The Power of Your Subconscious Mind
West Nyack, N.Y.: Parker, 1981, N.Y.: Bantam, 1982
Originally published in 1962

Die Macht Ihres Unterbewusstseins
München: Hugendubel, 2000

La puissance de votre subconscient
Genève: Ramón Keller, 1967

The Miracle of Mind Dynamics
New York: Prentice Hall, 1964

Miracle Power for Infinite Riches
West Nyack, N.Y.: Parker, 1972

The Amazing Laws of Cosmic Mind Power
West Nyack, N.Y.: Parker, 1973

Secrets of the I Ching
West Nyack, N.Y.: Parker, 1970

Think Yourself Rich
Use the Power of Your Subconscious Mind to Find True Wealth
Revised by Ian D. McMahan, Ph.D.
Paramus, NJ: Reward Books, 2001

Das Erfolgsbuch
Wie sie alles im Leben erreichen können
Hamburg: Heyne Verlag, 2002

Wahrheiten die ihr Leben verändern
Dr. Joseph Murphys Vermächtnis
München: Hugendubel, 1996

Murphy, Michael

The Future of the Body
Explorations into the Further Evolution of Human Nature
New York: Jeremy P. Tarcher/Putnam, 1992

Der Quanten-Mensch
München: Ludwig Verlag, 1996

Myers, Tony Pearce

The Soul of Creativity
Insights into the Creative Process
Novato, CA: New World Library, 1999

Myss, Caroline
The Creation of Health
The Emotional, Psychological, and Spiritual Responses that Promote Health and Healing
New York: Three Rivers Press, 1998

N

Naparstek, Belleruth
Your Sixth Sense
Unlocking the Power of Your Intuition
London: HarperCollins, 1998

Staying Well With Guided Imagery
New York: Warner Books, 1995

Narby, Jeremy
The Cosmic Serpent
DNA and the Origins of Knowledge
New York: J. P. Tarcher, 1999

Die Kosmische Schlange
Auf den Pfaden der Schamanen zu den Ursprüngen modernen Wissens
Stuttgart: Klett-Cotta, 2007

Nau, Erika
Self-Awareness Through Huna
Virginia Beach: Donning, 1981

Selbstbewusst durch Huna
Die magische Weisheit Hawaiis
2. Auflage
Basel: Sphinx Verlag, 1989

Neill, Alexander Sutherland
Neill! Neill! Orange-Peel!
New York: Hart Publishing Co., 1972

Neill! Neill! Birnenstiel
Berlin: Rowohlt, 1973

Summerhill
A Radical Approach to Child Rearing
New York: Hart Publishing, Reprint 1984
Originally published 1960

Theorie und Praxis der Antiautoritären Erziehung
Das Beispiel Summerhill
Berlin: Rowohlt Verlag, 1969

Summerhill School
A New View of Childhood
New York: St. Martin's Press
Reprint 1995

Das Prinzip Summerhill
Berlin: Rowohlt, 1971

Neuhaus, Heinrich

The Art of Piano Playing
London: Barrie & Jenkins, 1973
Reprinted 1997, 2001, 2002, 2006
First published in 1958

Neumann, Erich

The Great Mother
Princeton: Princeton University Press, 1955
(Bollingen Series)

Die Grosse Mutter
Die weiblichen Gestaltungen des Unterbewussten
Düsseldorf: Patmos Verlag, 2003

Newton, Michael
Life Between Lives
Hypnotherapy for Spiritual Regression
Woodbury, Minn.: Llewellyn Publications, 2006

Ni, Hua-Ching
I Ching
The Book of Changes and the Unchanging Truth
2nd edition
Santa Barbara: Seven Star Communications, 1999

Esoteric Tao The Ching
The Shrine of the Eternal Breath of Tao
Santa Monica: College of Tao and Traditional Chinese Healing, 1992

The Complete Works of Lao Tzu
Tao The Ching & Hua Hu Ching
Translation and Elucidation by Hua-Ching Ni
Santa Monica: Seven Star Communications, 1995

Nichols, Sallie
Jung and Tarot: An Archetypal Journey
New York: Red Wheel/Weiser, 1986

Die Psychologie des Tarot
Interlaken: Ansata Verlag, 1996

Nin, Anaïs
The Diary of Anaïs Nin (7 Volumes)
New York, 1966

Volume 1 (1931-1934)
New York: Harvest Books, 1969

Volume 2 (1934-1939)
New York: Harvest Books, 1970

O

O'Brian, Shirley
Child Pornography
2nd edition
New York: Kendall/Hunt, 1992

Odent, Michel
Birth Reborn
What Childbirth Should Be
London: Souvenir Press, 1994

The Scientification of Love
London: Free Association Books, 1999

Die Wurzeln der Liebe
Wie unsere wichtigsten Emotionen entstehen
Olten: Walter Verlag, 2001

Primal Health
Understanding the Critical Period Between Conception
and the First Birthday
London: Clairview Books, 2002
First Published in 1986 with Century Hutchinson in London

La Santé Primale
Paris: Payot, 1986

Die sanfte Geburt
Die Leboyer-Methode in der Praxis
Bergisch-Gladbach: Lübbe Verlag, 2001

The Functions of the Orgasms
The Highway to Transcendence
London: Pinter & Martin, 2009

Ollendorf-Reich, Ilse
Wilhelm Reich, A Personal Biography
New York, St. Martins Press, 1969

Wilhelm Reich
Vorwort von A.S. Neill
München, Kindler, 1975

Ong, Hean-Tatt
Amazing Scientific Basis of Feng-Shui
Kuala Lumpur: Eastern Dragon Press, 1997

Oppenheim, Lassa
International Law
4th Edition, by Sir Arnold D. McNair
New York, 1928

Ostrander, Sheila & Schroeder, Lynn
Superlearning 2000
New York: Delacorte Press, 1994

Superlearning
Die revolutionäre Lernmethode
München: Scherz Verlag, 1979

Supermemory
New York: Carroll & Graf, 1991

SuperMemory
Der Weg zum optimalen Gedächtnis
München: Goldmann, 1996

Ouspensky, Pyotr Demianovich
In Search of the Miraculous
New York: Mariner Books, 2001
First published in 1949

P

Papus
(Dr Gérard Encausse)
Traité de Magique Pratique
16e édition
St. Jean de Braye: Éditions Dangles, 1989

Patridge, Burgo
History of Orgies
New York, 1960

Pearce, John A. II and Robinson B. Jr.
Strategic Management
Formulation, Implementation and Control
Tenth Edition
New York: McGraw-Hill, 2007

Pearce Myers, Tony (Editor)
The Soul of Creativity
Insights into the Creative Process
Novato: New World Library, 1999

Pert, Candace B.
Molecules of Emotion
The Science Behind Mind-Body Medicine
New York: Scribner, 2003

Petrash, Jack
Understanding Waldorf Education
Teaching from the Inside Out
London: Floris Books, 2003

Phipson
Phipson on Evidence
13th ed., by John Huxley Buzzard
Richard May and M. N. Howard
London: Sweet & Maxwell, 1982

Plato
Complete Works
Ed. by John M. Cooper
New York: Hackett Publishing Company, 1997

Plummer, Kenneth
Pedophilia
Constructing a Sociological Baseline
in: in: Cook, M. and Howells, K. (Eds.):
Adult Sexual Interest in Children
Academic Press, London, 1980, pp. 220 ff.

Plutarch
Plutarch's Lives
The Dryden Translation
New York: Bantam Books, 2006

Ponder, Catherine
The Healing Secrets of the Ages
Marine del Rey: DeVorss, 1985

Porteous, Hedy S.
Sex and Identity
Your Child's Sexuality
Indianapolis: Bobbs-Merrill, 1972

Prescott, James W.
Affectional Bonding for the Prevention of Violent Behaviors
Neurobiological, Psychological and Religious/Spiritual Determinants
in: Hertzberg, L.J., Ostrum, G.F. and Field, J.R., (Eds.)

Violent Behavior
Vol. 1, Assessment & Intervention, Chapter Six
New York: PMA Publishing, 1990

Alienation of Affection
Psychology Today, December 1979

Body Pleasure and the Origins of Violence
Bulletin of the Atomic Scientists, 10-20 (1975)

Deprivation of Physical Affection as a Primary Process in the Development of Physical Violence A Comparative and Cross-Cultural Perspective,
in: David G. Gil, ed., Child Abuse and Violence
New York: Ams Press, 1979

Early somatosensory deprivation as an ontogenetic process in the abnormal development of the brain and behavior,
in: Medical Primatology, ed. by I.E. Goldsmith and J. Moor-Jankowski,
New York: S. Karger, 1971

Genital Mutilation of Children
Failure of Humanity and Humanism
Unprinted Essay (2005)
http://www.violence.de/prescott/letters/
CIRC_CONGRESS_MONTAGUE_9.30.05.html

Genital Pain vs. Genital Pleasure
Why the One and not the Other
The Truth Seeker, July/August 1989, pp. 14-21
http://www.violence.de/prescott/truthseeker/genpl.html

How Culture Shapes the Developing Brain and the Future of Humanity
A Brief Summary of the research which links brain abnormalities and violence to an absence of nurturing and bonding very early in childhood,
in: Touch the Future: Optimum Learning Relationships for Children & Adults
Spring 2002 (Ed. by Michael Mendizza)
Nevada City, CA, 2002

Invited Commentary: Central nervous system functioning in altered sensory environments,
in: M.H. Appley and R. Trumbull (Eds.), *Psychological Stress,*
New York: Appleton-Century Crofts, 1967

Our Two Cultural Brains: Neurointegrative and Neurodissociative
http://www.violence.de/prescott/letters/Our_Two_Cultural_Brains.pdf

Phylogenetic and ontogenetic aspects of human affectional development,
in: Progress in Sexology, Proceedings of the 1976 International Congress of Sexology,
ed. by R. Gemme & C.C. Wheeler
New York: Plenum Press, 1977

Prevention or Therapy and the Politics of Trust
Inspiring a New Human Agenda
in: *Psychotherapy and Politics International*
Volume 3(3), pp. 194-211
London: John Wiley, 2005

Sex and the Brain
Midcontinent & Eastern Regions, June 13-16, 2002
Big Rapids, MI: Society for Cross-Cultural Research,
32nd Annual Meeting, 2005
http://www.violence.de/archive.shtml

Sixteen Principles for Personal, Family and Global Peace
The Truth Seeker, March/April 1989
http://www.violence.de/prescott/letters/Sixteen_Principles.pdf

Somatosensory affectional deprivation (SAD) theory of drug and alcohol use,
in: Theories on Drug Abuse: Selected Contemporary Perspectives,
ed. by Dan J. Lettieri, Mollie Sayers and Helen Wallenstien Pearson,
NIDA Research Monograph 30, March 1980
Rockville, MD: National Institute on Drug Abuse,
Department of Health and Human Services, 1980

The Origins of Human Love and Violence
Pre- and Perinatal Psychology Journal
Volume 10, Number 3:
Spring 1996, pp. 143-188 The Origins of Love and Violence
Sensory Deprivation and the Developing Brain
Research and Prevention (DVD)
http://ttfuture.org/store/origins_orders

http://violence.de
http://ttfuture.org/violence
http://montagunocircpetition.org

Pritchard, Colin

The Child Abusers
New York: Open University Press, 2004

R

Radin, Dean

The Conscious Universe
The Scientific Truth of Psychic Phenomena
San Francisco: Harper & Row, 1997

Entangled Minds
Extrasensory Experiences in a Quantum Reality
New York: Paraview Pocket Books, 2006

Raknes, Ola

Wilhelm Reich and Orgonomy
Oslo: Universitetsforlaget, 1970

Wilhelm Reich und die Orgonomie
Eine Einführung in die Wissenschaft von der Lebensenergie
Frankfurt/M: Nexus, 1983

Randall, Neville
Life After Death
London: Robert Hale, 1999

Rank, Otto
Art and Artist
With Charles Francis Atkinson and Anaïs Nin
New York: W.W. Norton, 1989
Originally published in 1932

The Significance of Psychoanalysis for the Mental Sciences
New York: BiblioBazaar, 2009
First published in 1913

Rausky, Franklin
Mesmer ou la révolution thérapeutique
Paris, 1977

Redfield, James
The Tenth Insight
Holding the Vision
New York: Warner Books, 1996

The Celestine Prophecy
New York: Warner Books, 1995

Die Vision von Celestine
Berlin: Ullstein, 2004

Reich, Wilhelm
A Review of the Theories, dating from The 17th Century, on the Origin of Organic Life
by Arthur Hahn, Literature Assistant at the Institut für Sexualökonomischen Lebensforschung, Biologisches Laboratorium, Oslo, 1938
©1979 Mary Boyd Higgins as Director of the Wilhelm Reich Infant Trust
XEROX Copy from the Wilhelm Reich Museum

Children of the Future
On the Prevention of Sexual Pathology
New York: Farrar, Straus & Giroux, 1984
First published in 1950

CORE (Cosmic Orgone Engineering)
Part I, Space Ships, DOR and DROUGHT
©1984, Orgone Institute Press
XEROX Copy from the Wilhelm Reich Museum

Der Einbruch der sexuellen Zwangsmoral
Frankfurt/M: Fischer, 1981

Die Entdeckung des Orgons II
Der Krebs
Frankfurt/M: Fischer, 1981
Köln: Kiepenheuer & Witsch, 1984

Die Funktion des Orgasmus
Sexualökonomische Grundprobleme der biologischen Energie
Köln: Kiepenheuer & Witsch, 1987

Die Massenpsychologie des Faschismus
Frankfurt/M: Fischer, 1974

Die sexuelle Revolution
Frankfurt/M: Fischer, 1966

Early Writings 1
New York: Farrar, Straus & Giroux, 1975

Ether, God & Devil & Cosmic Superimposition
New York: Farrar, Straus & Giroux, 1972
Originally published in 1949

Frühe Schriften 1
Aus den Jahren 1920-1925
Frankfurt/M: Fischer, 1983

Frühe Schriften 2
Genitalität in der Theorie und Therapie der Neurose
Frankfurt/M: Fischer, 1985

Genitality in the Theory and Therapy of Neurosis
©1980 by Mary Boyd Higgins as Director of the Wilhelm Reich
Infant Trust

Leidenschaften der Jugend
Köln: Kiepenheuer & Witsch, 1984

L'irruption de la morale sexuelle
Paris: Payot, 1972

Menschen im Staat
Frankfurt/M: Nexus, 1982

People in Trouble
©1974 by Mary Boyd Higgins as Director of the Wilhelm Reich
Infant Trust

Record of a Friendship
The Correspondence of Wilhelm Reich and A. S. Neill
New York, Farrar, Straus & Giroux, 1981

Selected Writings
An Introduction to Orgonomy
New York: Farrar, Straus & Giroux, 1973

The Bioelectrical Investigation of Sexuality and Anxiety
New York: Farrar, Straus & Giroux, 1983
Originally published in 1935

The Bion Experiments
reprinted in *Selected Writings*
New York: Farrar, Straus & Giroux, 1973

The Cancer Biopathy (The Orgone, Vol. 2)
New York: Farrar, Straus & Giroux, 1973

The Function of the Orgasm (The Orgone, Vol. 1)
Orgone Institute Press, New York, 1942

The Invasion of Compulsory Sex Morality
New York: Farrar, Straus & Giroux, 1971
Originally published in 1932

The Leukemia Problem: Approach
©1951, Orgone Institute Press
Copyright Renewed 1979
XEROX Copy from the Wilhelm Reich Museum

The Mass Psychology of Fascism
New York: Farrar, Straus & Giroux, 1970
Originally published in 1933

The Orgone Energy Accumulator
Its Scientific and Medical Use
©1951, 1979, Orgone Institute Press
XEROX Copy from the Wilhelm Reich Museum

The Schizophrenic Split
©1945, 1949, 1972 by Mary Boyd Higgins as Director of the
Wilhelm Reich Infant Trust
XEROX Copy from the Wilhelm Reich Museum

The Sexual Revolution
©1945, 1962 by Mary Boyd Higgins as Director of the
Wilhelm Reich Infant Trust

Zeugnisse einer Freundschaft
Der Briefwechsel zwischen Wilhelm Reich und A.S.
Neill (1936-1957)
Köln: Kiepenheuer & Witsch, 1986

Reid, Daniel P.

The Tao of Health, Sex & Longevity
A Modern Practical Guide to the Ancient Way
New York: Simon & Schuster, 1989

Guarding the Three Treasures
The Chinese Way of Health
New York: Simon & Schuster, 1993

Renault, Mary

The Persian Boy
New York: Bantam Books, 1972

Reps, Paul

Zen Flesh, Zen Bones
Rutland: Tuttle Publishing, 1989

Rhodes, Richard

The Making of the Atomic Bomb
New York, Simon & Schuster, 1995

Richardson, Justin

Everything You Never Wanted Your Kids to Know About Sex
With Mark. A. Schuster
New York: Three Rivers Press, 2003

Richet, Charles

Metapsychical Phenomena
Methods and Observations
Kessinger Publishing Reprint Edition, 2004
Originally published in 1905

Riso, Don Richard & Hudson, Russ

The Wisdom of the Enneagram
The Complete Guide to Psychological and Spiritual Growth
For The Nine Personality Types
New York: Bantam Books, 1999

Robbins, Anthony

Awaken The Giant Within
New York: Simon & Schuster, 1991

Unlimited Power
The New Science of Personal Achievement
New York: Free Press, 1997

Roberts, Jane

The Nature of Personal Reality
New York: Amber-Allen Publishing, 1994
First published in 1974

Die Natur der Persönlichen Realität
Ein neues Bewusstsein als Quelle der Kreativität
München: Kailash Verlag, 2007

The Nature of the Psyche
Its Human Expression
New York, Amber-Allen Publishing, 1996
First published in 1979

Die Natur der Psyche
Ihr menschlicher Ausdruck in Kreativität, Liebe, Sexualität
Genf: Ariston Verlag, 1985

Die Natur der Psyche
Ihr menschlicher Ausdruck in Kreativität, Liebe, Sexualität
München: Kailash Verlag, 2008

Roman, Sanaya

Opening to Channel
How To Connect With Your Guide
New York: H.J. Kramer, 1987

Zum Höheren Selbst Erwachen
Das Herz dem Bewusstsein des Lichts öffnen
Genf: Ansata Verlag, 2003

Rosen, Sydney (Ed.)
My Voice Will Go With You
The Teaching Tales of Milton H. Erickson
New York: Norton & Co., 1991

Rosenbaum, Julius
The Plague of Lust
New York: Frederick Publications, 1955

Rossman, Parker
Sexual Experiences between Men and Boys
New York, 1976

Rothschild & Wolf
Children of the Counterculture
New York: Garden City, 1976

Rousseau, Jean-Jacques
Émile ou de l'Éducation, 1762
Reprint, Paris: Garnier, 1964

The Social Contract
And Later Political Writings
Cambridge, MA.: Cambridge University Press, 1997

Rudhyar, Dane
Astrology of Personality
A Reformulation of Astrological Concepts and Ideals in
Terms of Contemporary Psychology and Philosophy
New York: Aurora Press, 1990

An Astrological Triptych
Gifts of the Spirit, The Way Through, and The Illumined Road
New York: Aurora Press, 1991

Astrological Mandala
New York: Vintage Books, 1994

L'astrologie de la transformation
Paris: Rocher, 1984

Ruiz, Don Miguel

The Four Agreements
A Practical Guide to Personal Freedom
San Rafael, CA: Amber Allen Publishing, 1997

The Mastery of Love
A Practical Guide to the Art of Relationship
San Rafael, CA: Amber Allen Publishing, 1999

The Voice of Knowledge
A Practical Guide to Inner Peace
With Janet Mills
San Rafael, CA: Amber Allen Publishing, 2004

Ruperti, Alexander

Cycles of Becoming
The Planetary Pattern of Growth
New York: CRCS Publications, 1978

La roue de l'expérience individuelle
Paris: Librairie de Médicis, 1991

Rush, Florence

The Best Kept Secret
Sexual Abuse of Children
New Jersey: Prentice-Hall, 1980

Das bestgehütete Geheimnis
Sexueller Kindesmissbrauch
Berlin: Sub-Rosa Frauenverlag, 1984

S

Saint-Simon, Claude-Henri de
De la réorganisation de la société européenne
Avec Auguste Thierry
Paris, 1814
Lausanne: Centre de Recherches Européennes, 1967

Salas, Floyd
Tatoo the Wicked Cross
New York: Grove Press, 1967

Salomé, Jacques
Si je m'écoutais, je m'entendrais
Avec Sylvie Galland
Paris: Éditions de l'Homme, 1990

Sandfort, Theo
The Sexual Aspect of Pedophile Relations
The Experience of Twenty-five Boys
Amsterdam: Pan/Spartacus, 1982

SantoPietro, Nancy
Feng Shui, Harmony by Design
How to Create a Beautiful and Harmonious Home,
New York: Putnam-Berkeley, 1996

Satinover, Jeffrey
Homosexuality and the Politics of Truth
New York: Baker Books, 1996

The Quantum Brain
New York: Wiley & Sons, 2001

Satprem
Sri Aurobindo ou l'aventure de la conscience
Paris: Buchet/Castel, 1970

Scarro A. M., Jr. (Ed.)
Male Rape
New York: Ams Press, 1982

Schérer, René
Co-ire
Album systématique de l'enfance
Avec Guy Hocquenghem
Recherches No. 22
Paris: E.S.F., 1976

Émile perverti, ou des rapports entre l'éducation et la sexualité
Paris: Robert Laffont, 1974
Paris, Désordres, 2006
Nouvelle Édition

Le corps interdit
Avec Georges Lapassade
Paris: E.S.F., 1976

Une érotique puérile
Paris: Éditions Galilée, 1978

Schlipp, Paul A. (Ed.)
Albert Einstein
Philosopher-Scientist
New York: Open Court Publishing, 1988

Schonberg, Harold
The Great Pianists
From Mozart to the Present
New York: Simon and Schuster (Fireside), 2006
Originally published in 1963

Schrenck-Notzing, Albert von
Phenomena of Materialization
A Contribution to the Investigation of Mediumistic Teleplastics
Perspectives in Psychical Research
New York: Kegan Paul, 1920

Schultes, Richard Evans, et al.
Plants of the Gods
Their Sacred, Healing, and Hallucinogenic Powers
New York: Healing Arts Press
2nd edition, 2002

Die Pflanzen der Götter
Die magischen Kräfte der Rausch- und Giftgewächse
München: AT Verlag, 1998

Schumacher, E.F.
Small is Beautiful
Economics as if People Mattered
San Francisco: Harper Perennial, 1989

Schwartz, Andrew E.
Guided Imagery for Groups
Fifty Visualizations That Promote Relaxation, Problem-Solving, Creativity, and Well-Being
Whole Person Associates, 1995

Senf, Bernd
Die Wiederentdeckung des Lebendigen
Aachen: Omega, 2003
Erstmals veröffentlicht 1996 mit Zweitausendeins Verlag in Frankfurt/M

Nach Reich: Neue Forschungen zur Orgonenergie
Sexualökonomie / Die Entdeckung der Orgonenergie
Herausgegeben zusammen mit Professor James DeMeo,
Ashland, Oregon, USA
Frankfurt/M: Zweitausendeins Verlag, 1997

Sepper, Dennis L.
Goethe Contra Newton
Polemics and the Project of a New Science of Color
Cambridge: Cambridge University Press, 1988

Shalabi, Ahmad
Islam
Cairo, 1970

Sharaf, Myron
Fury on Earth
A Biography of Wilhelm Reich
London: André Deutsch, 1983

Wilhelm Reich
Der heilige Zorn des Lebendigen
Berlin: Simon & Leutner, 1994

Sheldrake, Rupert
A New Science of Life
The Hypothesis of Morphic Resonance
Rochester: Park Street Press, 1995

Das Schöpferische Universum
Die Theorie des morphogenetischen Feldes
Neue und erweiterte Auflage
Berlin: Ullstein, 2009

Sher, Barbara & Gottlieb, Annie
Wishcraft
How to Get What You Really Want
2nd edition
New York: Ballantine Books, 2003

Shone, Ronald
Creative Visualization
Using Imagery and Imagination for Self-Transformation
New York: Destiny Books, 1998

Simonton, O. Carl et al.
Getting Well Again
Los Angeles: Tarcher, 1978

Singer, June
Androgyny
New York: Doubleday Dell, 1976

Smith, C. Michael
Jung and Shamanism in Dialogue
London: Trafford Publishing, 2007

Spiller, Jan
Astrology for the Soul
New York: Bantam, 1997

Spock, Benjamin
Dr. Spock's Baby and Child Care
8th Edition
New York: Pocket Books, 2004

Säuglings- und Kinderpflege
Berlin: Ullstein, 1986

Spretnak, Charlene
Green Politics
Rochester, VT: Inner Traditions, 1986

Stein, Robert M.
Redeeming the Inner Child in Marriage and Therapy
in: Reclaiming the Inner Child
ed. by Jeremiah Abrams
New York: Tarcher/Putnam, 1990, 261 ff.

Steiner, Rudolf
Theosophy
An Introduction to the Spiritual Processes in Human Life
and in the Cosmos
New York: Anthroposophic Press, 1994

Die Erziehung des Kindes
Dornach: Rudolf Steiner Verlag, 2003
First published in 1907

Stekel, Wilhelm
Auto-Eroticism
A Psychiatric Study of Onanism and Neurosis
Republished, London: Paul Kegan, 2004

Patterns of Psychosexual Infantilism
New York, 1959 (reprint edition)

Psychosexueller Infantilismus
Die seelischen Kinderkrankheiten der Erwachsenen
Berlin: Urban & Schwarzenberg, 1922

Sadism and Masochism
New York: W.W. Norton & Co., 1953

Sex and Dreams
The Language of Dreams
Republished
New York: University Press of the Pacific, 2003

Störungen des Trieb- und Affektlebens
Bände I & II
Berlin: Urban & Schwarzenberg, 1921

Stiene, Bronwen & Frans
The Reiki Sourcebook
New York: O Books, 2003

The Japanese Art of Reiki
A Practical Guide to Self-Healing
New York: O Books, 2005

Stone, Hal & Stone, Sidra
Embracing Our Selves
The Voice Dialogue Manual
San Rafael, CA: New World Library, 1989

Du bist viele
Das 100fache Selbst und seine Entdeckung durch
die Voice-Dialogue Methode
München: Heyne Verlag, 1994

Strassman, Rick
DMT: The Spirit Molecule
A doctor's revolutionary research into the biology of near-death
and mystical experiences
Rochester: Park Street Press, 2001

Stratenwerth, Günter
Schweizerisches Strafrecht
Besonderer Teil II, 3. Aufl.
Bern: Stämpfli, 1984

Sun Tzu (Sun Tsu)
The Art of War
Special Edition
New York: El Paso Norte Press, 2007

Die Kunst des Krieges
Hamburg: Nikol Verlag, 2008

Suryani, Luh Ketut & Jensen, Gorden D.
The Balinese People
A Reinvestigation of Character
New York: Oxford University Press, 1993

Sutherland
Statutory Construction
Ed. By Sands, 4th Edition
London, 1975

Sweeny/Oliver/Leech
The International Legal System
Cases and Materials
2nd Edition
Minneola, N.Y.: Foundation Press, 1981

Symonds, John Addington
A Problem in Greek Ethics
New York: M.S.G. House, 1971

Szasz, Thomas
The Myth of Mental Illness
New York: Harper & Row, 1984

T

Talbot, Michael
The Holographic Universe
New York: HarperCollins, 1992

Das holographische Universum
Die Welt in neuer Dimension
München: Droemer Knaur, 1994

Tansley, David V.
Chakras, Rays and Radionics
London: Daniel Company Ltd., 1984

Targ, Russell & Katra, Jane
Miracles of Mind
Exploring Nonlocal Consciousness and Spiritual Healing
Novato, CA: New World Library, 1999

Tarnas, Richard
Cosmos and Psyche
Intimations of a New World View
New York: Plume, 2007

The Passion of the Western Mind
Understanding the Ideas that have Shaped Our World View
New York: Ballantine Books, 1993

Tart, Charles T.
Altered States of Consciousness
A Book of Readings
Hoboken, N.J.: Wiley & Sons, 1969

Tatar, Maria M.
Spellbound: Studies on Mesmerism and Literature
Princeton, N.Y., 1978

Tchouang-tseu
Oeuvre complète
Paris: Gallimard/Unesco, 1969

Temple, Robert
The Sirius Mystery
New Scientific Evidence of Alien Contact 5000 Years Ago
Rochester: Destiny Books, 1998

Textor, R. B.
A Cross-Cultural Summary
New Haven, Human Relations Area Files (HRAF)
Press, 1967

The Advent of Great Awakening
A Course in Miracles
Text Workbook and Manual for Teachers
New York: New Christian Church of Full Endeavor, 2007

The Tibetan Book of the Dead
The Great Liberation through Hearing in the Bardo
Translated with commentary by Francesca
Fremantle & Chögyam Trungpa
Boston: Shambhala Dragon Editions, 1975

The Ultimate Picasso
New York: Harry N. Abrams, 2000

Thorsson, Edred
Futhark
A Handbook of Rune Magic
San Francisco: Weiser Books, 1984

Tiller, William A.
Conscious Acts of Creation
The Emergence of a New Physics
Associated Producers, 2004 (DVD)

Psychoenergetic Science
New York: Pavior, 2007

Conscious Acts of Creation
New York: Pavior, 2001

Tischner, Rudolf
F.A. Mesmer
München, 1928

Todaro-Franceschi, Vidette
The Enigma of Energy
Where Science and Religion Converge
New York: Crossroad Publishing, 1991

Toffler, Alvin
Powershift
Knowledge, Wealth, and Violence at the Edge of the 21st Century
New York: Bantam, 1991

Revolutionary Wealth
How it will be created and how it will change our lives
New York: Broadway Business, 2007

The Third Wave
New York: Bantam, 1984

Tolle, Eckhart
The Power of Now
A Guide to Spiritual Enlightenment
Novato, CA: New World Library, 2004

Jetzt! Die Kraft der Gegenwart
Ein Leitfaden zum spirituellen Erwachen
Bielefeld: Kamphausen Verlag, 2000

A New Earth
Awakening to Your Life's Purpose
New York: Michael Joseph (Penguin), 2005

Eine neue Erde
Bewusstseinssprung anstelle von Selbstzerstörung
München: Goldmann, 2005

Too, Lillian
Feng Shui
Kuala Lumpur: Konsep Books, 1994

U

Unlawful Sex
Offences, Victims and Offenders in the Criminal Justice System of England and Wales
The Report of the Howard League Working Party
London: Waterloo Publishers Ltd., 1985

V

Van Gelder, Dora
The Real World of Fairies
A First-Person Account
Wheaton: Quest Books, 1999
First published in 1977

Vanguard, Thorkil
Phallós
A Symbol and its History in the Male World
New York: International Universities Press, 2001

Villoldo, Alberto
Healing States
A Journey Into the World of Spiritual Healing and Shamanism
With Stanley Krippner
New York: Simon & Schuster (Fireside), 1987

Dance of the Four Winds
Secrets of the Inca Medicine Wheel
With Eric Jendresen
Rochester: Destiny Books, 1995

Die Macht der vier Winde
Eine Reise ins Reich der Schamanen
München: Goldmann, 2009

Shaman, Healer, Sage
How to Heal Yourself and Others with the Energy Medicine
of the Americas
New York: Harmony, 2000

Hüter des alten Wissens
Schamanisches Heilen im Medizinrad
Darmstadt: Schirner Verlag, 2007

Healing the Luminous Body
The Way of the Shaman with Dr. Alberto Villoldo
DVD, Sacred Mysteries Productions, 2004

Mending The Past And Healing The Future with Soul Retrieval
New York: Hay House, 2005

Seelenrückholung: die Vergangenheit schamanistisch erkunden
Die Zukunft heilen
München, Goldmann, 2006

Vitebsky, Piers

The Shaman
Voyages of the Soul, Trance, Ecstasy and Healing from
Siberia to the Amazon
New York: Duncan Baird Publishers, 2001
Originally published in 1995

Von Riezler, Sigmund

Geschichte der Hexenprozesse in Bayern
Stuttgart: Magnus Verlag, 1983

W

Walker & Walker
The English Legal System
6th Edition, by R.J. Walker
London: Butterworths, 1985

Ward, Elizabeth
Father-Daughter Rape
New York: Grove Press, 1985

Watts, Alan W.
The Way of Zen
New York: Vintage Books, 1999

This Is It
And Other Essays on Zen and Spiritual Experience
New York: Vintage, 1973

Wee Chow Hou
The 36 Strategies of the Chinese
Adapting Ancient Chinese Wisdom to the Business World
New York: Addison-Wesley, 2007

Weiss, Jess E.
The Vestibule
New York: Ashley Books, 1979

West's Encyclopedia of American Law
Second Edition
New York: Gale Group, 2008

Wharton
Wharton's Criminal Law
14th ed. by Charles E. Torcia
Vol. II, §§99-282
Rochester, New York: The Lawyers Cooperative Publishing Co., 1979

What the Bleep Do We Know!?
See Arntz, William

Whiteman
Digest of International Law
Vol. 6
Washington, D.C.: Department of State Publication 8350, 1968

Whitfield, Charles L.
Healing the Child Within
Deerfield Beach, Fl: Health Communications, 1987

Whiting, Beatrice B.
Children of Six Cultures
A Psycho-Cultural Analysis
Cambridge: Harvard University Press, 1975

Wiener, Jon
Gimme Some Truth: The John Lennon FBI Files
Los Angeles: University of California Press, 1999

Wilber, Ken
Sex, Ecology, Spirituality
The Spirit of Evolution
Boston: Shambhala, 2000

Quantum Questions
Mystical Writings of The World's Greatest Physicists
Boston: Shambhala, 2001

Wild, Leon D.
The Runes Workbook
A Step-by-Step Guide to Learning the Wisdom of the Staves
San Diego: Thunder Bay Press, 2004

Wilhelm Helmut
The Wilhelm Lectures on the Book of Changes
Princeton: Princeton University Press, 1995

Wilhelm, Richard
The I Ching or Book of Changes
With C. Baynes
3rd Edition, Bollingen Series XIX
Princeton, NJ: Princeton University Press, 1967

Williams, Strephon Kaplan
Dreams and Spiritual Growth
With Patricia H. Berne and Louis M. Savary
New York: Paulist Press, 1984

Durch Traumarbeit zum eigenen Selbst
Die Jung-Senoi Methode
Interlaken: Ansata Verlag, 1987

Dream Cards
Understand Your Dreams and Enrich Your Life
New York: Simon & Schuster (Fireside), 1991

Wing, R. L.
The I Ching Workbook
Garden City, N.Y.: Doubleday, 1984

Das Arbeitsbuch zum I Ching
Mit Chinesischen Orakel Münzen
München: Goldmann, 2004

Het I Tjing Werkboek
Baarn: Bigot & Van Rossum, 1986

Woerly, Franz
Esprit Guide
Entretiens avec Karlfried Dürckheim
Paris: Albin Michel, 1985

Wolf, Fred Alan
Taking the Quantum Leap
The New Physics for Nonscientists
New York: Harper & Row, 1989

Der Quantensprung ist keine Hexerei
Frankfurt/M: Fischer Verlag, 1990

Parallel Universes
New York: Simon & Schuster, 1990

The Dreaming Universe
A Mind-Expanding Journey into the Realm Where Psyche and Physics Meet
New York: Touchstone, 1995

The Eagle's Quest
A Physicist Finds the Scientific Truth At the Heart of the Shamanic World
New York: Touchstone, 1997

Die Physik der Träume
Frankfurt/M: DTV Verlag, 1997

Mind into Matter
A New Alchemy of Science and Spirit
New York: Moment Point Press, 2000

Words and Phrases Legally Defined
Ed. By John Saunders
2nd Edition
London: Butterworths, 1969

Wydra, Nancilee
Feng Shui
The Book of Cures
Lincolnwood: Contemporary Books, 1996

Y

Yang, Jwing-Ming
Qigong, The Secret of Youth
Da Mo's Muscle/Tendon Changing and Marrow/Brain Washing Classics
Boston, Mass.: YMAA Publication Center, 2000

The Root of Chinese Qigong
Secrets for Health, Longevity, & Enlightenment
Roslindale, MA: YMAA Publication Center, 1997

Yates, Alayne
Sex Without Shame
Encouraging the Child's Healthy Sexual Development
New York, 1978
Republished Internet Edition

Yeats, William Butler
Irish Fairy and Folk Tales
New York: Modern Library, 2003

Mythologies
New York: Simon & Schuster, 1998
Author Copyright 1959, Renewed 1987 by Anne Yeats

Ywahoo, Dhyani
Voices of Our Ancestors
Cherokee Teachings from the Wisdom Fire
New York: Shambhala, 1987

Am Feuer der Weisheit
Lehren der Cherokee Indianer
Zürich: Theseus Verlag, 1988

Z

Znamenski, Andrei A.
Shamanism
Critical Concepts in Sociology
New York: Routledge, 2004

Zinker, Joseph
Se créer par la Gestalt
Montréal: Les Éditions de l'Homme, 1981

Zukav, Gary
The Dancing Wu Li Masters
An Overview of the New Physics
New York: HarperOne, 2001

Die tanzenden Wu Li Meister
Der östliche Pfad zum Verständnis der modernen Physik
Vom Quantensprung zum schwarzen Loch
Berlin: Rowohlt, 2000

Zweig, Stefan
Die Heilung durch den Geist
Mesmer, Mary Baker-Eddy, Freud
Frankfurt/M: Fischer Verlag, 1982
Originally published in 1931

Zyman, Sergio
The End of Marketing as We Know It
New York: HarperCollins, 2000

Das Ende der Marketing Mythen
Erfolgsrezepte des Aya-Cola für Umsatz und Profit
Berlin: Econ Verlag, 2000

Anthropologists discovered that their gaze was a tool of domination and that their discipline was not only a child of colonialism, it also served the colonial cause through its own practices. The unbiased and supra-cultural language of the observer was actually a colonial discourse and a form of domination.
– Jeremy Narby, *The Cosmic Serpent (1998/2003)*, p. 14.

FROM THE SAME AUTHOR

A Bibliography

You can search publications from here:
http://ipublica.com/books/

For audio books and music, you can start here:
http://ipublica.com/audio/

All paperbacks, audio downloads, audio book compact discs, music downloads and music compact discs, as well as Kindle books, are referenced on the site.

For free podcasts search iTunes under my author name.

For quoting my publications, please use the following form:
Pierre F. Walter, [Title]: [Subtitle], Newark: Sirius-C Media Galaxy LLC, 2011

Web Presence

Pierre F. Walter on the Web

Sites

http://authoryourlife.com

http://ipublica.com

http://ipublica.net

http://ipublica.org

http://ipublica.tv

Video Channels

http://youtube.com/user/ipublica

http://youtube.com/user/authoryourlife

http://vimeo.com/pierrefwalter/channels

http://ipublica.blip.tv/

http://authoryourlife.blip.tv/

http://emosexuality.blip.tv/

http://pierrefwalter.blip.tv/

www.ingramcontent.com/pod-product-compliance
Lightning Source LLC
Chambersburg PA
CBHW030938180526
45163CB00002B/618